Lecture Notes in Bioinformatics 11347

Subseries of Lecture Notes in Computer Science

More information about this series at http://www.springer.com/series/5381

Luis Kowada · Daniel de Oliveira (Eds.)

Advances in Bioinformatics and Computational Biology

12th Brazilian Symposium on Bioinformatics, BSB 2019
Fortaleza, Brazil, October 7–10, 2019
Revised Selected Papers

 Springer

Editors
Luis Kowada
Fluminense Federal University
Niterói, Brazil

Daniel de Oliveira
Fluminense Federal University
Niterói, Brazil

ISSN 0302-9743   ISSN 1611-3349 (electronic)
Lecture Notes in Bioinformatics
ISBN 978-3-030-46416-5   ISBN 978-3-030-46417-2 (eBook)
https://doi.org/10.1007/978-3-030-46417-2

LNCS Sublibrary: SL8 – Bioinformatics

This Springer imprint is published by the registered company Springer Nature Switzerland AG
The registered company address is: Gewerbestrasse 11, 6330 Cham, Switzerland

Preface

This volume contains the papers presented at the 12th Brazilian Symposium on Bioinformatics (BSB 2019), which took place during October 7–10, 2019, in Niterói, Brazil. BSB is an international conference that covers all aspects of bioinformatics and computational biology. The event was organized by the special interest group in Computational Biology of the Brazilian Computer Society (SBC), which has been the organizer of BSB for the past several years. The BSB series started in 2005. In the period 2002–2004, its name was Brazilian Workshop on Bioinformatics (WOB).

As in previous editions, BSB 2019 had an international Program Committee (PC) of 31 members. After a rigorous review process by the PC, 12 papers were accepted to be orally presented at the event (8 full papers and 4 short papers), and are printed in this volume. All papers were reviewed by at least two independent reviewers. We believe that this volume represents a fine contribution to current research in computational biology and bioinformatics, as well as in molecular biology. In addition to the technical presentations, BSB 2019 featured keynote talks from Daniel de Oliveira (Universidade Federal Fluminense) and Werner Treptow (Universidade de Brasília).

BSB 2019 was made possible by the dedication and work of many people and organizations. We would like to express our sincere thanks to all PC members, as well as to the external reviewers. Their names are listed in the pages that follow. We are also grateful to the local organizers and volunteers for their valuable help; the sponsors for making the event financially viable; and Springer for agreeing to publish this volume. Finally, we would like to thank all authors for their time and effort in submitting their work and the invited speakers for having accepted our invitation.

November 2019

Luis Kowada
Daniel de Oliveira

Organization

Conference Chair

Sergio Lifschitz   Pontifícia Universidade Católica do Rio de Janeiro, Brazil

Program Chairs

Luis Antonio Brasil Kowada   Universidade Federal Fluminense, Brazil

Daniel de Oliveira   Universidade Federal Fluminense, Brazil

Steering Committee

João Carlos Setubal   Universidade de São Paulo, Brazil
Luis Antonio Kowada   Universidade Federal Fluminense, Brazil
Natália Florencio Martins   Empresa Brasileira de Pesquisa Agropecuária, Brazil
Ronnie Alves   Instituto Tecnológico Vale, Brazil
Sérgio Vale Aguiar Campos   Universidade Federal de Minas Gerais, Brazil
Tainá Raiol   Fundação Oswaldo Cruz, Brazil
Waldeyr Mendes   Instituto Federal de Goiás, Brazil

Program Committee

Alexandre Paschoal   Universidade Federal Tecnológica do Paraná, Brazil
André C. Ponce de Leon F. de Carvalho   Universidade de São Paulo, Brazil
André Kashiwabara   Universidade Federal Tecnológica do Paraná, Brazil
Annie Chateau   Université de Montpellier, France
César Manuel Vargas Benítez   Universidade Federal Tecnológica do Paraná, Brazil
Daniel de Oliveira   Universidade Federal Fluminense, Brazil
Fabrício Martins Lopes   Universidade Federal Tecnológica do Paraná, Brazil
Felipe Louza   Universidade de São Paulo, Brazil
Fernando Luís Barroso Da Silva   Universidade de São Paulo, Brazil
Guilherme Pimentel Telles   Universidade Estadual de Campinas, Brazil
Ivan G. Costa   RWTH Aachen University, Germany
Jefferson Morais   Universidade Federal do Pará, Brazil
João Carlos Setubal   Universidade de São Paulo, Brazil
Kleber Padovani de Souza   Universidade Federal do Pará, Brazil
Laurent Bréhélin   Université de Montpellier, France

Luciana Montera	Instituto Tecnológico Vale, Brazil
Luciano Antonio Digiampietri	Universidade de São Paulo, Brazil
Luís Felipe Ignácio Cunha	Universidade Federal do Rio de Janeiro, Brazil
Luis Antonio Kowada	Universidade Federal Fluminense, Brazil
Marcilio De Souto	Université d'Orléans, France
Marcio Dorn	Universidade Federal do Rio Grande do Sul, Brazil
Maria Emilia Telles Walter	Universidade de Brasilia, Brazil
Mariana Recamonde-Mendoza	Universidade Federal do Rio Grande do Sul, Brazil
Marilia Braga	Bielefeld University, Germany
Nalvo Franco de Almeida Jr.	Universidade Federal de Mato Grosso do Sul, Brazil
Natália Florencio Martins	Empresa Brasileira de Pesquisa Agropecuária, Brazil
Rommel Ramos	Universidade Federal do Pará, Brazil
Ronnie Alves	Instituto Tecnológico Vale, Brazil
Said Sadique Adi	Universidade Federal de Mato Grosso do Sul, Brazil
Sérgio Vale Aguiar Campos	Universidade Federal de Minas Gerais, Brazil
Sergio Lifschitzs	Pontifícia Universidade Católica do Rio de Janeiro, Brazil
Sergio Pantano	Institut Pasteur de Montevideo, Uruguay
Tainá Raiol	Fundação Oswaldo Cruz, Brazil
Waldeyr Mendes Cordeiro da Silva	Instituto Federal de Goiás, Brazil
Zanoni Dias	Universidade Estadual de Campinas, Brazil

Sponsors

Sociedade Brasileira de Computação (SBC)
Coordenação de Aperfeiçoamento de Pessoal de Nível Superior (CAPES)
Springer Verlag

Contents

Extended Abstracts

Full Papers

On Clustering Validation
in Metagenomics Sequence Binning

Paulo Oliveira[1], Kleber Padovani[1], and Ronnie Alves[1,2]

[1] Computer Science Graduate Program, Federal University of Pará, Belém, Brazil
p.paulo.f.oliveira@gmail.com, kleber.padovani@gmail.com
[2] Instituto Tecnológico Vale, Belém, Brazil
ronnie.alves@itv.org

Abstract. In clustering, one of the most challenging aspects is the validation, whose objective is to evaluate how good a clustering solution is. Sequence binning is a clustering task on metagenomic data analysis. The sequence clustering challenge is essentially putting together sequences belonging to the same genome. As a clustering problem it requires proper use of validation criteria of the discovered partitions. In sequence binning, the concepts of precision and recall, and F-measure index (external validation) are normally used as benchmark. However, on practice, information about the (sub) optimal number of cluster is unknown, so these metrics might be biased to an overestimated "ground truth". In the case of sequence binning analysis, where the reference information about genomes is not available, how to evaluate the quality of bins resulting from a clustering solution? To answer this question we empirically study both quantitative (internal indexes) and qualitative aspects (biological soundness) while evaluating clustering solutions on the sequence binning problem. Our experimental study indicates that the number of clusters, estimated by binning algorithms, do not have as much impact on the quality of bins by means of biological soundness of the discovered clusters. The quality of the sub-optimal bins (greater than 90%) were identified in both rich and poor clustering partitions. Qualitative validation is essential for proper evaluation of a sequence binning solution, generating bins with sub-optimal quality. Internal indexes can only be used in compliance with qualitative ones as a trade-off between the number of partitions and biological soundness of its respective bins.

Keywords: Validation · Clustering · Unsupervised · Binning · Metagenomics

1 Introduction

Metagenomic analysis seeks the recovery of genomes from the mixture of sequence fragments (metagenome). In this context, the clustering of DNA fragments into a corresponding taxonomic group is called binning (Fig. 1), where each sequence is allocated to a group that ideally represents only the fragments

© Springer Nature Switzerland AG 2020
L. Kowada and D. de Oliveira (Eds.): BSB 2019, LNBI 11347, pp. 3–15, 2020.
https://doi.org/10.1007/978-3-030-46417-2_1

belonging to a particular taxon or genome [1]. Binning is critical for the reconstruction of genomes because it can improve other steps of the analysis, such as assembly, and, in addition, provide knowledge about the genetic material of the metagenome as a whole. In the context of metagenomics analysis, binning means clustering.

The difficulty of recovering genomes is analogous to the assembly of a puzzle, which is composed of several pieces in which it is not known how many are similar and how they fit to solve the problem. Binning can make the analysis more robust by grouping such pieces, that is, sequences with similar biological patterns. Currently, there are two categories of binning methods: taxonomic and genome binning [2]. The first type classifies the DNA fragments when comparing them with a genome reference in a database, while the latter performs an unsupervised classification, applying clustering techniques using extracted features (k-mers frequency and GC content) [1], solely based on the sequences that make up the sample under analysis.

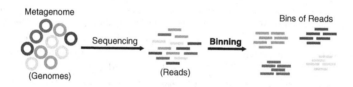

Fig. 1. Sequence (reads) binning workflow.

Existing genome binning methods use a variety of clustering algorithms such as k-means [3,4] and Expectation-Maximization [5]. Although they are based on unsupervised classification methods, these methods validate their solutions from reference-dependent metrics and concepts like precision, recall and F-measure. Since genome binning is applied in samples/data set whose genome content is unknown, thus, the number of clusters is unknown. Cluster validation based on supervised metrics is not coherent. Genome binning methods usually employed supervised metrics.

Some binning tools are based on the use of genetic markers from a reference database [6,7]. These tools, when using genetic markers, adopt, albeit indirectly, the use of quality measures to improve the clustering result, but ignore the validation. Quality assessment in terms of completeness and contamination of bins resulting from clustering is critical and can be estimated based on the frequency of single-copy genetic markers [8,9].

We evaluate the use of quality measures as a validation criteria in the unsupervised binning methods along with internal clustering indexes. These indexes do not require a priori knowledge of the data, and have extensive practical applications in information retrieval, text analysis, image processing and analysis, bioinformatics, and other data mining application domains. The quality of the clustering results is usually evaluated internally from two aspects, intra-cluster compression and inter-cluster separation. In order to better understand the internal indexes in the validation of the binning of short metagenomic sequences, three well-known indexes established in the literature were explored in this work: Index Silhouette [10], Index Davies-Bouldin [11] and Index Calinski-Harabasz [12].

This work proposes in the validation of clusters of sequences to adopt the use of measures and statistics based on the quantitative and qualitative aspects of clustering, allowing to evaluate the utility of the bins resulting from metagenomics sequence binning.

2   On Clustering Metagenomic Data

Clustering plays a key role in metagenomic analysis. Sequence clustering methods can answer many of the fundamental questions by grouping similar sequences into groups of genes or genomes. In addition, clustering allows one to deal directly with the challenges of metagenomics [13]. Recently, there was the emergence and improvement of several clustering techniques applied in metagenomic analysis. Some examples are Metacluster [3], Maxbin [6] and Metabat [14], which specifically address sequence grouping through distinct approaches. However, validating the outcome of these techniques is still a critical issue. The resulting partitions are commonly validated by statistical measures, which often do not take into account whether these partitions correspond to the actual biological structures that make up the data set.

Although clustering is a tool that offers great advantages in data processing, in the metagenomic there are still some problems that make clustering not having the same potential. And this is directly linked to the difficulty in dealing with the data. Metagenomic data are usually of high dimensionality, having more variables than samples, having a lot of noise and missing data. These characteristics pose problems for most clustering methods.

Given the difficulty of this analysis, there is no consensus regarding the best clustering, feature selection, or better distance function that can be used in a binning analysis in the metagenomic [15]. Therefore, it is common to use several methods to analyze a given set of data and choose the most appropriate result based on a curated analysis by a domain specialist, in this case, a biologist. In this way, the analysis can become subjective by making clusters overestimated, while disregarding clusters that can result in relevant conclusions if analyzed by another perspective.

It is important to note that although the use of biological knowledge is necessary in the final interpretation of a clustering analysis, it does not mean that they can replace an unsupervised validation step, in which the significance of clusters is analyzed individually from implicit data extracted strictly from the analyzed data set.

Clustering depends on the existence of a structural pattern in the data. In metagenomic binning analysis, the expected structures refer to the clustering of sequences belonging to a single genome. However, most clustering algorithms return a cluster, even in the absence of a real structure, leaving the researcher to identify the lack of significance of the returned results. This can lead to a lack of compliance between partitioning and the distribution of implicit data, which is the type of case that can be detected using cluster validation techniques.

3   Clusters Validation

The validation of clusters generally allows to evaluate the quality of the partitions. However, there are several clustering methods that can produce different results for the same data set [16]. With this in mind, the validation allows the selection of a set of clusters from the same method executed with different parameters, the number of clusters, or from different clustering methods.

The main goal of cluster validation is to find the best partitioning that fits the distribution of the characteristics extracted from the data. Two common criteria are used to evaluate and select an optimal clustering scheme:

- Compactness: The items in each cluster should be as close as possible to each other. A widely used measure of compaction is variance.
- Separation: Clusters should be well separated from each other. There are three widely used approaches to measure distance between two clusters.
  - Distance between items closest to clusters;
  - Distance between the most distant items;
  - Distance between the centroids of the clusters.

Clustering solutions are commonly evaluated by the lens of distinct validation indexes, namely external, internal and relative ones [17]. External validation methods, as its name suggests, evaluate clustering based on external data as a reference. The internal validation methods are based on quantitative metrics that are calculated from the data set itself and the clustering scheme. Relative validation has as principle the comparison of different clustering schemes. Several clustering algorithms are executed several times with different parameters for the same data set, whose objective is to find the best clustering strategy from the different results obtained.

4   Quantitative Validation (Internal Indexes)

The internal indexes allow to evaluate the results of clustering based on quantities and characteristics inherent to the data set. A brief definition of the internal indexes used in this study is presented below.

- **Silhouette:** Its evaluation is based on the compactness and separation criteria of clusters. In an interval of $[-1,1]$, an average Silhouette value close to 1 indicates a better overall quality of the clustering result. On the other hand, a value close to zero indicates poor clustering quality, and negative values indicate poorly defined clusters [10].

- **Davies-Bouldin:** It measures the average similarity between each cluster and its most similar. Small values correspond to clusters that are compact and have centers distant from each other. Therefore, a minimum value represents a better cluster quality [11].
- **Calinski-Harabasz:** It is based on the maximization of the intercluster dispersion and the minimization of the intracluster dispersion. The maximum value represents a better clustering quality [12].

One of the main advantages of using internal validation indexes is the possibility of estimating the optimal number of clusters [16]. In this case, the optimal clustering solution is found from a series of clustering solutions under different numbers of clusters. However, it is not always feasible to run series of clusters for a given task or set of data. Some clustering applications use their own methods to estimate the number of clusters. In these cases, the number of clusters (k) estimated does not mean an optimum subsequent clustering, because using different clustering algorithms for the same k (optimal or not) can result in different clustering solutions from the same data set. The binning algorithms studied here fit into these cases, and therefore the internal indexes in question are used as a measure to evaluate the ratio of the number of clusters, estimated by the binning tools, to the quality scores (biological soundness) of the binning results.

5 Qualitative Validation

In the study [18], one of the topics covered deals with the use of external biological information to improve and evaluate the quality of association patterns of discovered genes. They state that quality results from conducting a biological evaluation before assures the biological soundness of the patterns found. It has also been observed that researchers are more concerned to scalability issues than quality ones. These observations are valid for clustering metagenomic data, because when it comes to quality assessment, its utility remains ignored.

Recent binning tools, such as DAS tool [15], already use CheckM [8] to generate bins quality estimates based on genetic markers, thus allowing comparison of the quality of bins of different methods of binning. The DAS tool is a consensual binning tool, and uses the quality to choose the best bins from various tools to then generate its consensus. CheckM is a tool that generates the estimation of the completeness and contamination of the genome, statistics that allow to estimate the quality of genomes. Besides quality, biological soundness contribute to the significance of what is a well-formed bin; other statistics such as N50, fraction of the genome, and number of contigs play critical role to quality assessment. Genome assembly is not taken into account by DAS tool.

In our work the N50 statistics, fraction of the genome and number of contigs are obtained with the MetaQUAST [19] tool. It is a reference-based method that identifies errors in a genome assembly relative to the reference genomes. MetaQUAST is a modification of the QUAST [20] tool, a genome assembly validation tool that calculates alignments of contigs assembled to a single reference genome. CheckM and MetaQUAST are genome evaluation tools, and we believe that the quality of the bins can also be verified by analyzing the assembly of the bin itself. Thus, by biological soundness we mean bins having a good trade-off between genome assembly and content.

6   Experimental Study

6.1   Data Sets

The data sets are composed of 24 samples, already used in [4, 21, 22]. They are synthetic metagenomes generated using the MetaSim [23] tool. These data sets are divided into three groups: S, L, and R. The S and L data sets are composed of paired-end short reads (approximately 80 bp), generated according to Illumina profile error with error rate of 1%. The data sets L are composed of genomes of two species and are oriented to the evaluation of binning algorithms in relation to different abundance rates between two species. The S samples have a greater variation in the number of species, abundance rate and phylogenetic distance. The R samples are composed of Roche 454 single-end long reads (approximately 700 bp) and 1% sequencing error rate. It is important to note that each generated sample presents intrinsic data complexity; high complexity samples are those ones having high species diversity in the microbial community under study. Low complexity samples are composed of a microbial community having low species diversity. Samples of higher complexity poses more challenge to sequence clustering [1]. It is the case of some of the R and S samples. In the L samples, there are only two species, thus, the complexity is linked to the variation of the abundance of its associated species. More details about the simulated data sets can be found in the supplementary material at gitlab.com/binning_validation/supp-data.

6.2   Methods

The workflow (Fig. 2) for clustering validation of sequence bins is defined as follows:

1) Sequence clustering: Sequence binning was performed using the MetaProb [21], MetaCluster [3] and MBBC [5] tools. These tools has been selected since they are the state-of-the-art in sequence binning, and they also estimate the number of clusters.
2) Quantitative validation: Once having the clustering results, all internal indexes were calculated using in-house Python scripts. The inputs for the scripts are the sequence samples and the list of cluster indexes. The DNA sequences are strand strings {A, C, G, T} and they require to be transformed to feature vectors before the calculation of its corresponding internal indexes:
   - Sequence vectorization: Sequence are decomposed to tetra-nucleotides frequencies, following the CONCOCT [24] approach. CONCOCT relies on PCA for data dimensionality reduction. Thus, PCA is applied over short reads, so there is less information to be kept in a single short sequence.

In [25] it is demonstrated that by reducing the dimension of the data there is an increase of the silhouette score for biological data (DNA sequences), and they empirically suggest $D = 2$ for a better representation of the data for k-means clustering.

- Clustering indexation: Each sequence has been indexed following binning results, and next the list of predicted labels for each sample is generated for the calculation of the internal indexes; the silhouette.py script returns the scores and the graph, whereas DB.py and CH.py return only the score. Detailed scripts, scores and charts can be found at gitlab.com/binning_validation/supp-data.

3) Qualitative validation: In the next stage, all bins were evaluated by CheckM and MetaQUAST tools. Before qualitative evaluation each bin generated by the binning tools was assembled with Megahit [26]. Sequence assembly is mandatory for finding gene markers and proper functional annotation. CheckM provides two indexes to evaluate genome quality: i) the completeness and ii) contamination statistics. These indexes allowed us to obtain a quality score similar as the one devised in [27], where quality was defined as (*completeness* − *5* × *contamination*). The averages, the maximum qualities and the counts of bins with more than 90% of quality were recovered. The other qualitative indexes based on assembled bins such as N50, genome fraction and number of contigs were obtained by MetaQUAST. Then, the results of the averages of N50, the number of contigs, and the count of bins with more than 90% of genome fraction were recovered.

4) Outcomes: All quantitative and qualitative statistics are summarized. More details can be found at gitlab.com/binning_validation/supp-data.

7 Results and Discussion

7.1 Quantitative Validation

The evaluation of the number of clusters detected by each tool showed that there is a difficulty in estimating the number of clusters. The Table 1 shows that there were few cases where the correct K value was found, recall rate of 21.74%, less than 1/4 of the total. The lowest values were obtained for more complex paired-end (S) samples (8.33%), and the highest ones for paired-end (L) samples

Table 1. The number of clusters (k) calculated by each binning tool.

Binning tools																								
Sample	R1	R2	R3	R4	R5	R6	R7	R8	R9	S1	S2	S3	S4	S5	S6	S7	S8	S9	L1	L2	L3	L4	L5	L6
MetaCluster	1	1	**2**	**2**	**2**	2	**3**	7	4	9	15	30	9	27	34	119	19	139	13	25	17	29	11	11
MetaProb	10	4	5	3	4	6	8	6	14	**2**	1	8	6	7	8	7	6	35	4	**2**	4	**2**	**2**	5
MBBC		1	1	1	1	1	2	**3**	2	4	1	1	1	1	1	**3**	−	−	−	1	**2**	**2**	**2**	**2**	**2**
Real value	2	2	2	2	2	3	3	3	6	2	2	2	2	3	3	5	5	15	2	2	2	2	2	2

[1]The correct K values are depicted in bold.

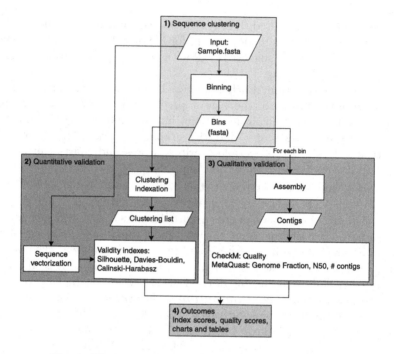

Fig. 2. Workflow of the proposed clustering validation.

(44.44%). Indeed, the sample complexity impact strongly on the selection of the number of clusters and, consequently, the final clustering solution.

The MBBC tool presented better results, lower error rate (rr) regarding the ability to estimate the real value of K, Table 2. However, it will be shown that the best results of clustering were not necessarily achieved by discovering the best number of clusters. The error rate was obtained by the following equation:

$$rr_j = \sum_1^m \left| n - \sum_1^i EV_i - RV_i \right|,$$

where m = group of samples {R,S,L}, n = number of samples of each group m, i = samples of each group m, j = binning tool, EV = estimated number of clusters, RV = real value of number of clusters.

Table 2. The overall (rr) scoring values for each binning tool. The lowest the value the better the estimation of the number of partitions.

Samples	MetaCluster	MetaProb	MBBC
R	0.33	13	3.5
S	45.83	7.5	2.67
L	47	3.5	0.5
Total	93.17	24	**6.67**

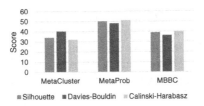

Fig. 3. The overall scoring values of the binning tools for each internal validity index.

When the internal validation index scores were analyzed, Fig. 3, it was observed that, in general, the clusters that presented the best quality were the ones calculated by the MetaProb tool. However, for the less complex samples (L) the best results were achieved with MBBC. The score of the indexes was based on the ascending order of the best result for each binner, as follows: the binner with the best score adds 3 points, the second best 2 points and the last 1 point. Next, we add up all scores for each sample, and so the binner who got the best results scored more points. Sequence binning are dependent on the input data and selected binning tool, consequently the internal indexes can vary dramatically for the same data set. The consensus approach introduced in [15] may help to deal with this problem, although it still depends on the proper selection of the binning tools used for the consensus.

7.2   Qualitative Validation

The qualitative validation is based on the evaluation of biological soundness of the assembled bins. So, the results will refer to both genome assembly and content. Binning tools are either focused on contigs (sequence having thousands of bp) or reads (sequence having hundreds of bp). The latter poses more computational challenge. Cluster validation by means of evaluating the genome content must take into account large sequences, thus assembly bins are critical for proper assessment of genome content [7–9].

We selected Megahit [26] for getting quality statistics of the bin assembled. Megahit is one of the most used metagenome assembler. Ideally, good genome assembly would produce few contigs with high N50 and high genome fraction values [28].

The results for genome assembly through the N50 and number of contigs indicators can be seen in the Tables A and B in the supplemental material. The bins obtained by the MBBC returned, on average, larger contigs. However, it also produced more contigs, among the three tools, it might be an indication of genome redundancy (more about genome redundancy in [29,30]). In contrast, the MetaCluster presented smaller number of contigs, however, smaller sequences indicating a possible greater fragmentation (more about genome fragmentation in [31,32]).

Regarding the quality of the contigs, the fraction of the genome is the best metric representing both the quality of the assembly and the contents of the contigs. Table 3 shows the results of the number of bins with genome fraction greater than 90%. It is considered that a low genome fraction rate (<85%) may indicate a significant difference between the reference genome and the sequenced one [33]. Of the total of 258 bins generated by MetaProb, 138 obtained genome fraction greater than 90%, thus a recall rate of 53.49%. MetaProb do not have the best recall rates (correct number of clusters), but it has provided high quality bins. The quality results are summarized by averaging the quality of bins along each sample (Table C and D in supplement material).

Table 3. The fraction of genome higher than 90%.

Samples	MetaCluster	MetaProb	MBBC
R	1	**42**	6
S	0	**81**	1
L	0	**15**	2
Total	1	**138**	9

Figure 4 shows binners providing quality bins between 70% and 90%, and suboptimal bins with quality above 90%. MetaProb, therefore, generated more bins with medium and high qualities. However, when analyzing the proportion of total generated bins to the bins recovered with the mentioned qualities (Fig. 5), it can be noticed that the MBBC returned more expressive results.

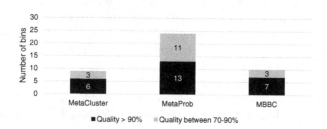

Fig. 4. Number of bins with medium/high quality score.

Our cluster validation proposal raises quantitative and qualitative aspects of binning that are not addressed by DAS tool, since the goal there is to obtain the consensus of the bins regardless of its quality. The statistical divergences w.r.t. some of the results highlight how difficult clustering is. However, bringing these statistics into the analysis can have a considerable impact on metagenomics sequence binning, since the quality of the recovered bins are critical for the utility of these recovered genomes [27, 34].

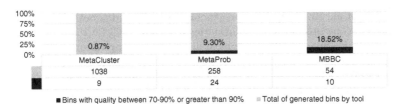

Fig. 5. Ratio of number of bins generated by bins recovered with medium/high quality.

8 Conclusion

Clustering is a difficult task on many domains. Given that the ground truth is usually unknown, there are always have issues regarding the quality of the induced clusters. Qualitative analysis is mandatory in the sense of bringing real utility of those patterns on the domain. Genome binning tools are essentially clustering strategies, and thus, biological soundness of the resulting bins must be checked by some external biological information. Our experimental evaluation study reinforces the needs to explore qualitative information for proper assessment of genome bins. Genome assembly and content statistics of bins can be used as an additional validation measures tightly linked to the main goal which is recovering individual genomes from metagenomic data.

References

1. Mande, S.S.: Classification of metagenomic sequences: methods and challenges. Brief. Bioinform. **13**, 669–681 (2012)
2. Sedlar, K.: Bioinformatics strategies for taxonomy independent binning and visualization of sequences in shotgun metagenomics. Comput. Struct. Biotechnol. J. **15**, 48–55 (2017)
3. Wang, Y., et al.: MetaCluster 5.0: a two-round binning approach for metagenomic data for low-abundance species in a noisy sample. Bioinformatics **28**(18), i356–i362 (2012)
4. Vinh, L., et al.: A two-phase binning algorithm using l-mer frequency on groups of non-overlapping reads. Algorithms Mol. Biol. **10**, 2 (2015). https://doi.org/10.1186/s13015-014-0030-4
5. Wang, Y., et al.: MBBC: an efficient approach for metagenomic binning based on clustering. BMC Bioinform. **16**, 36 (2015)
6. Wu, Y., et al.: MaxBin: an automated binning method to recover individual genomes from metagenomes using an expectation-maximization algorithm. Microbiome **2**, 26 (2014). https://doi.org/10.1186/2049-2618-2-26
7. Lin, H., Yu-Chieh, L.: Accurate binning of metagenomic contigs via automated clustering sequences using information of genomic signatures and marker genes. Sci. Rep. **6**, 24175 (2016)
8. Parks, D., et al.: CheckM: assessing the quality of microbial genomes recovered from isolates, single cells, and metagenomes. Genome Res. **25**, 1043–1055 (2015)

9. Simão, F., et al.: BUSCO: assessing genome assembly and annotation completeness with single-copy orthologs. Bioinformatics **31**, 1367–4803 (2015)
10. Rousseeuw, P.: Silhouettes: a graphical aid to the interpretation and validation of cluster analysis. J. Comput. Appl. Math. **20**, 53–65 (1987)
11. Davies, D.L., Bouldin, D.W.: A cluster separation measure. Trans. Pattern Anal. Mach. Intell. **1**(2), 224–227 (1979)
12. Calinski, T., Harabasz, J.: A dendrite method for cluster analysis. Commun. Stat. **3**(1), 1–27 (1974)
13. Li, W., et al.: Ultrafast clustering algorithms for metagenomic sequence analysis. Brief. Bioinform. **13**(6), 656–668 (2012)
14. Kang, D., Froula, J., Egan, R., Wang, Z.: MetaBAT, an efficient tool for accurately reconstructing single genomes from complex microbial communities. PeerJ **3**, e1165 (2015)
15. Sieber, C., et al.: Recovery of genomes from metagenomes via a dereplication, aggregation and scoring strategy. Nat. Microbiol. **3**, 836–843 (2018)
16. Van Craenendonck, T., Blockeel, H.: Using internal validity measures to compare clustering algorithms. Benelearn (2015)
17. Legány, C., Juhász, S., Babos, A.: Cluster validity measurement techniques. In: Proceedings of the 5th WSEAS International Conference on Artificial Intelligence (2006)
18. Alves, R., Rodriguez-Baena, D.S., Aguilar-Ruiz, J.S.: Gene association analysis: a survey of frequent pattern mining from gene expression data. Brief. Bioinform. **11**(2), 210–224 (2010)
19. Mikheenko, A., Saveliev, V., Gurevich, A.: MetaQUAST: evaluation of metagenome assemblies. Bioinformatics **32**(7), 1088–1090 (2016)
20. Gurevich, A., Saveliev, V., Vyahhi, N., Tesler, G.: QUAST: quality assessment tool for genome assemblies. Bioinformatics **29**(8), 1072–1075 (2013)
21. Girotto, S., Pizzi, C., Comin, M.: MetaProb: accurate metagenomic reads binning based on probabilistic sequence signatures. Bioinformatics **32**(17), i567–i575 (2016)
22. Reyes, P., Villegas, C.: An empirical comparison of EM and K-means algorithms for binning metagenomics datasets. Ingeniare. Rev. Chil. Ing. **26**, 20–27 (2018)
23. Richter, D.C., et al.: MetaSim: a sequencing simulator for genomics and metagenomics. PLoS ONE **3**, e3373 (2018)
24. Alneberg, J., Bjarnason, B.S., De Bruijn, I., Schirmer, M., Quick, J., Ijaz, U.Z., et al.: Binning metagenomic contigs by coverage and composition. Nat. Methods **11**(11), 1144–1146 (2014)
25. Baridam, B.B., Ali, M.M.: An investigation of K-means clustering to high and multi-dimensional biological data. Kybernetes **42**(4), 614–627 (2013)
26. Li, D., et al.: MEGAHIT v1.0: a fast and scalable metagenome assembler driven by advanced methodologies and community practices. Methods **102**, 3–11 (2016)
27. Parks, D., et al.: Recovery of nearly 8,000 metagenome-assembled genomes substantially expands the tree of life. Nat. Microbiol. **2**, 1533–1542 (2017)
28. Khan, A.R., et al.: A comprehensive study of de novo genome assemblers: current challenges and future prospective. Evol. Bioinform. Online **14** (2018)
29. Krakauer, D.C., Plotkin, J.B.: Redundancy, antiredundancy, and the robustness of genomes. Proc. Nat. Acad. Sci. U.S.A. **99**(3), 1405–1409 (2002)
30. Chen, H.W., et al.: Predicting genome-wide redundancy using machine learning. BMC Evol. Biol. **10**, 1471–2148 (2010)
31. Klassen, J.L., Currie, C.R.: Gene fragmentation in bacterial draft genomes: extent, consequences and mitigation. BMC Genom. **13**, 14 (2012)

32. Poptsova, M.S., et al.: Non-random DNA fragmentation in next-generation sequencing. Sci. Rep. **4**, 4532 (2014)
33. Mikheenko, A., Prjibelski, A., Saveliev, V., Antipov, D., Gurevich, A.: Versatile genome assembly evaluation with QUAST-LG. Bioinformatics **34**(13), i142–i150 (2018)
34. Sangwan, N., Xia, F., Gilbert, J.: Recovering complete and draft population genomes from metagenome datasets. Microbiome **04**(1), 2049–2618 (2016)

Genome Assembly Using Reinforcement Learning

Roberto Xavier[1], Kleber Padovani de Souza[1], Annie Chateau[3], and Ronnie Alves[1,2]

[1] Federal University of Pará, Belém, PA, Brazil
rbxjunior@gmail.com, kleber.padovani@gmail.com, alvesrco@gmail.com
[2] Instituto Tecnológico Vale, Belém, PA, Brazil
[3] University of Montpellier, Montpellier, France
chateau@lirmm.fr

Abstract. Reinforcement learning (RL) aims to build intelligent agents able to optimally act after the training process to solve a given goal task in an autonomous and non-deterministic fashion. It has been successfully employed in several areas; however, few RL-based approaches related to genome assembly have been found, especially when considering real input datasets. *De novo* genome assembly is a crucial step in a number of genome projects, but due to its high complexity, the outcome of state-of-art assemblers is still insufficient to assist researchers in answering all their scientific questions properly. Hence, the development of better assembler is desirable and perhaps necessary, and preliminary studies suggest that RL has the potential to solve this computational task. In this sense, this paper presents an empirical analysis to evaluate this hypothesis, particularly in higher scale, through performance assessment along with time and space complexity analysis of a theoretical approach to the problem of assembly proposed by [2] using the RL algorithm *Q-learning*. Our analysis shows that, although space and time complexities are limiting scale issues, RL is shown as a viable alternative for solving the DNA fragment assembly problem.

Keywords: Reinforcement learning · Q-learning · Machine learning · Genome assembly

1 Introduction

One of the most important tasks in genome projects is to obtain the complete genome sequence of an organism, which is acquired by sequencing and assembly technologies working together [4]. However, the sequencing method used by current sequencers has a chemical limitation that prevents the reading of

This study was financed in part by the Coordenação de Aperfeiçoamento de Pessoal de Nível Superior - Brasil (Capes).
Finance Codes: 88882.460068/2019-01 and 88882.445004/2019-01.

© Springer Nature Switzerland AG 2020
L. Kowada and D. de Oliveira (Eds.): BSB 2019, LNBI 11347, pp. 16–28, 2020.
https://doi.org/10.1007/978-3-030-46417-2_2

the whole genome at once, thus being able to read only short fragments of DNA molecules [3]. Therefore, several sequencers use a technique named *Shotgun Sequencing*, which is based on the random cutting of the original DNA molecules into smaller fragments which can be read by sequencers—giving rise to textual sequences usually referred to as *reads*.

Genome assembly process, on the other hand, attempts to combine such reads in order to recover the sequence of nucleotides compounding the original DNA molecule [8]. The computational task that underlies genome assembly is called DNA fragment assembly (DFA) problem, which is considered an NP-hard combinatorial challenge—therefore, there is no algorithm capable of solving it deterministically in a reasonable (polynomial) time [13]. Heuristic strategies have been commonly used to solve the problem, producing reasonable but sub-optimal solutions.

Thereby, better assemblers able to recover genome sequences even closer to the real content of corresponding DNA molecules are still required [9,12]. In this context, many researchers have investigated the application of various computational techniques searching for better solutions to the DFA problem, including machine learning (ML) techniques [13].

Reinforcement learning (RL) is a branch of ML whose algorithms aim to train an intelligent agent to learn specific behaviours considering series of actions taken by the agent and observing how these actions modify the environment [14]. Rewards received as actions are taken by agents are generally the basis of learning for RL algorithms. In other words, RL uses a different learning modelling and solves problems by the execution of successive actions taken by an intelligent agent, which is not explicitly supervised, but receives rewards during the learning process that ideally promote the necessary knowledge to find the optimal solution, that corresponds to the set of actions which produces the maximum accumulated reward.

ML approaches for DFA problem are still rare, especially those using RL [13]. Bocicor et al. [2] proposed a pioneer RL-based model to genome assembly, using *Q-learning* algorithm for training an agent able to find the correct order of a set of reads allowing later assembly of them to recover the original genome. This approach, although successful, is extremely theoretical, with experiments that used few very short reads originated from genomes whose lengths are insignificant compared to real genomes.

Hence, the analysis of the scalability and, consequently, suitability of the proposed model for the resolution of real cases of the DFA problem becomes relevant. Hereupon, the objective of this paper is to expand the cited approach, considering larger input datasets, aiming to evaluate the viability of building an autonomous RL-based assembler to deal with scenarios closer to the real cases faced by genomics.

2 Background

2.1 ML-Based Genome Assembly

Current assemblers play an important role in genomic projects, although their solutions are generally sub-optimal. However, handling sequences and tuning the different parameters of these assemblers properly are not trivial tasks for all users. There are several aspects to consider in configuring tools for different types of sequences and organisms to obtain relevant results. We believe the application of ML techniques can build assemblers capable of autonomously deal with genome assembly issues, making this process simpler.

Indeed, ML has been among the approaches used to try to find better solutions to the DFA problem. Souza et al. present in [13] a broad review about ML-based approaches closely related to genome assembly and they classified the reviewed works in three different categories, named pre-assembly, post-assembly, and auto-assembly.

Pre-assembly category consists of approaches where ML has been applied before assembly process (for instance, in sequencing error detection); approaches where ML was applied after assembly process—as in assembly evaluation – were grouped in [13] in the post-assembly category; finally, auto-assembly grouped the approaches where ML techniques were applied during the assembly itself.

In this last category, there are works that combined heuristic strategies traditionally used by popular assemblers with ML algorithms, in order to optimize their performances, and there are also works that exclusively employed ML to assemble reads, as the RL-based approach proposed in [2], described below.

2.2 RL-Based Auto-assembly

According to [14], a RL problem consists of a task to be performed, an agent— responsible for performing this task—, and an environment—with which the agent interact to achieve a goal. The environment is modeled by states, which represent the various possible configurations of this environment, and actions, that can change the current state of the environment and are taken by the agent.

After performing an action in a certain state, the agent can receive a reward, previously defined during the environment modeling. Ideally, these rewards must be sufficient for the agent to learn the best action to be taken in any state of the environment—that is, the action that maximizes the sum of all rewards received, starting from an initial state and reaching an optimal final state. In this process of seeking the maximum expected future rewards, the agent has to take into account how good immediate actions are in the long term, since an action can generate low immediate reward, but guiding the agent in the future to states that generate high rewards, as well as the opposite may be true.

Bocicor et al. [2] addressed the DFA problem by building a RL model able to predict the correct order of a set of n reads so that they can be assembled afterwards to recover the original DNA sequence. In this modelling, each state represents one possible arrangement of fragments. Any state that represents a permutation with n fragments—i.e., an arrangement with n fragments without replacement—is assigned as a final state; additionally, there is only one initial state in the state space, which represents the absence of reads (no read).

Each read corresponds to an action and each action implies a transition from a state to another, so that the corresponding read is added to the arrangement of reads represented by the current state whenever an action is taken (for instance, after taking the action corresponding to some read identified as A in a state that represents the ordered arrangement with reads C and B, the current state of the environment becomes the state which represents the ordered arrangement for reads C, B, and A).

Rewards for each action taken vary according to the type of the state that is consequently reached. Actions leading to non-final states—including all states representing arrangements with less than n reads and/or arrangements with replacements—receive a constant small reward; whereas other actions—those leading to final states—receive greater rewards that correspond the overlaps between all consecutive pairs of reads in the corresponding permutation, as detailed below.

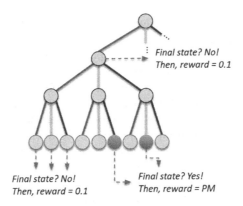

Fig. 1. Representation of the proposed state space for DNA problem.

Figure 1 illustrates the state space modelled by [2]. As mentioned, the set of actions available to the agent is composed of reads to be ordered, represented in the figure by state transitions of different colors. After taking an action—that is, after choosing one out n reads—, the environment reaches another state (final or non-final). Actions leading to non-final states receive a reward of 0.1 in this figure, and the overlap sum for the remaining actions are denoted as PM (standing for performance measure), defined by $PM(F) = \sum_{j=1}^{n-1} w(F_j, F_{j+1})$, where F represents an ordered set of textual sequences, F_i corresponds to the ith text, and $w(a, b)$ indicates the similarity between sequences a and b, measured by the local alignment algorithm Smith-Waterman [11].

Q-learning [15] was the algorithm used to learn the optimal policy for the agent, using ϵ-*Greedy* strategy, which is presented in Algorithm 1 and discussed next.

Algorithm 1: Q-learning algorithm (adapted from [14])

---

1 Initialize Q_{table} arbitrarily;
2 **for** $e \leftarrow 1$ **to** N **do**
3   Initialize S with an initial state;
4   **while** S *is not stoppable* **do**
5   Choose an action A according to a policy based on Q_{table};
6   Take action A and observe corresponding reward R and next state S';
7          $Q_{table}[S, A] \leftarrow$
         $Q_{table}[S, A] + \alpha(R + \gamma \max_{a \in actions}(Q_{table}[S', a]) - Q_{table}[S, A])$;
8          $S \leftarrow S'$;
9   **end**
10 **end**

The foundation of *Q-learning* is an important data structure, called *Q-table*, that is a table containing estimate accumulated future rewards for all possible states. *Q-table* values will guide the agent policy to choose actions during the training process. ϵ-*Greedy* strategy is used in this choice in order to balance *exploration-exploitation* of state space, so that the probabilities of the agent to choose random actions are greater at the beginning of the learning stage—thus allowing a broader investigation of state space (exploration), and over time, the probabilities progressively fall and, consequently, the agent begins to take actions based on the experiences acquired, observing the values contained in the *Q-table* (exploitation).

As stated in Algorithm 1, the agent takes many actions during a number (N) of cycles, referred to as episodes, that starts in an initial state and, after successive actions taken, finishes at a stoppable state—corresponding in [2] to any state (final or non-final) representing arrangements with n reads.

After each action a taken by the agent in a state s, the element in *Q-table* corresponding to (s, a) is updated by using *Bellman Equation* [1], which is presented at line 7 of Algorithm 1 and consists of discount factor (γ) and learning rate (α), that respectively determine a discount to be applied to future actions and the immediate impact of each table update.

According to the proposed modelling, the action space A corresponds to n, where n is the number of reads to be ordered, and the state space size S is defined as $\frac{n^{n+1}-1}{n-1}$. Thus, the larger the number of DNA fragments to be ordered, the larger the set of actions and the state space, and consequently the higher the *Q-table* dimension.

Among the experiments performed, this RL-based model was tested considering a small segment of a bacterium genome composed of 25 nucleotides, which gave rise to 10 synthetic reads with 8 *base pairs* (*bp*). After 4 million episodes, the proposed assembler produced optimal ordering for these reads, allowing the recovering of the original genome. Despite the optimal result, as mentioned, the input datasets used are very small, presenting number and size of reads so far from real cases faced by genomics—$25bp$ does not even reach 0.001% the size of the complete genome where this segment was extracted from.

3 Materials and Methods

3.1 Read Simulation

Experiments were conducted on 18 samples synthetically produced, 9 of them coming from a $50bp$ segment (hereafter referred to as microgenome) of the real genome of a bacterium and 9 from another microgenome, extracted from the reverse complement of the same real genome and consisting of $25bp$. All microgenomes and codes used in this paper, as well as supplementary files, can be downloaded at our repository files[1].

By using a read simulation script presented in [10], 9 samples were generated synthetically for each microgenome—each sample contains 10, 20 or 30 single-end reads (error-free) with the same size (8, 10 or $15bp$).

Since the read generation script tries to follow the same procedure adopted by sequencers, reads come from the same direction of the genome strand but also from the opposite direction (reverse complement). In order to circumvent this situation and to reach the closure of any gaps in all the samples of both microgenomes used, the reads corresponding to the reverse complement generated by the script were converted and later they were manually adapted so that all the corresponding microgenomes were covered in all cases.

3.2 Overlapping Scores and Ground Truth Acquisition

The reads of each sample were aligned on the respective microgenome so that the expected orders were obtained and, subsequently, the corresponding PM value was calculated. This value was used to evaluate the algorithm, considering the ratio of the obtained PM for the order given by the model to the expected PM, here termed relative PM.

In order to optimize the performance of experiments—since overlap computation is a high complexity task, the calculation of the overlapping scores between all pairs of reads for each sample was previously performed. The average time for calculating all the possible overlap pairs was 8.6 s. Thus, before starting the experiments, the overlapping values, expressed by the Smith-Waterman algorithm, for all possible combinations of reads were already calculated, stored and available for query.

Since the expected output of the assembly process is a textual sequence that ideally corresponds to the original genome of a set of reads, a complementary measure to the PM was used for the performance analysis of the RL-based

---

[1] http://doi.org/10.17605/OSF.IO/8C4ST.

assembler from another perspective. This metric intends to evaluate the quality of the sequence—denominated consensus sequence—that will be produced by ordering reads according to the order found by the assembler compared to the expected sequence (the microgenome).

Considering that the simulated reads are error-free, the consensus sequence for any given ordering of reads considered only exact matches between subsequent reads. Also taking into account that the PM value for a given order of reads is exactly equal to the PM obtained by these reads in the reverse order (i.e., PM(A, B, C) = PM(C, B, A)), it was also considered as a possible assembly the consensus sequence corresponding to the inverse order of the ordering achieved by the assembler. In other words, in order to obtain the distance between the corresponding microgenome and the output sequence of the RL-based assembler, the distance between the microgenome and the forward consensus sequence as well as the distance between the microgenome and the backward consensus sequence were calculated, prevailing the shorter one.

Levenstein's distance [5] was used to calculate the distance between any pair of textual sequences, assuming the same impact (equals to 1) for each change, removal or insertion of character required to transform a textual sequence into the other.

3.3 RL-Based Auto-Assembly Implementation

Although our object of study is the work introduced by Bocicor et al. [2], the source codes used in [2] were not publicly available. Therefore, except for the Levenstein's distance algorithm and the read generation scripts, all other algorithms have been implemented in-house and are available online.

Following the investigated approach, the Smith-Waterman algorithm was used in all experiments to calculate overlaps—with $match = 1.0$, $mismatch = -0.33$, and $gap = -1.33$—and Q-learning algorithm using ϵ-Greedy strategy was used to train the intelligent agent—with $\alpha = 0.85$, $\gamma = 0.90$, and $\epsilon = 1.0$ (applying ϵ-decay of $1/episodes$ after each episode and minimum ϵ of 0.1).

Training and validation were performed in triplicate for all 18 samples, considering increasing numbers of episodes, with 1, 2, 3, 4, and 5 million episodes in each experiment. Training time, number of states explored by Q-learning, and the corresponding performances of the models were extracted from each execution[2].

Finally, to assist the behavior analysis of the state space exploration and time for training (both as a function of the number of episodes), an extra replicate experiment with all the samples was carried out during 10 million episodes, following the same methodology cited above.

---

[2] All experiments were carried out on a cluster with an Intel(R) Xeon(R) CPU E5-4650 v3 at 2.10 GHz, with 384 Cores and 2 TB of RAM.

4   Results and Discussion

Table 1 shows the average relative PM (i.e., $\frac{obtainedPM}{expectedPM}$) among 3 executions after 5 million episodes for each of the 18 experiments. Similarly, Table 1 presents the Levenstein's distance for each of them.

Table 1. Average relative PM and Levenstein's distance across 3 runs after 5 million episodes

	Performance measure						Levenstein's distance					
	$25bp$-long microgenome			$50bp$-long microgenome			$25bp$-long			$50bp$-long		
	10	20	30	10	20	30	10	20	30	10	20	30
$8bp$	100,00%	82,41%	73,31%	106,64%	86,09%	70,91%	0	53	114	26	51	117
$10bp$	100,01%	88,70%	76,06%	100,59%	79,13%	67,32%	24	70	126	30	80	156
$15bp$	100,00%	93,82%	89,73%	100,00%	83,92%	72,58%	0	70	157	0	112	235

The first experiment—which corresponds exactly to the experiment performed by [2], containing 10 reads of $8bp$ from the microgenome of $25bp$—is one of the few cases where relative PM and distance optimally converged after training (PM equals to 100% and distance equals to 0).

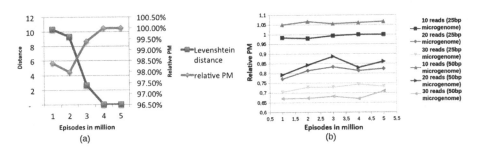

Fig. 2. (a) Relative PM and distance for the experiment with 10 reads of 8 derived from a $25bp$-long microgenome. (b) Relative PM for the experiments with reads of $8bp$

For further analysis, the progression of these metrics throughout the episodes for this experiment—hereafter referred to as a control experiment—are shown in Fig. 2(a). In agreement with the statement of Bocicor et al. in [2], it is possible to observe that the convergence of results occurs with 4 million training episodes for both metrics, which means that the agent retrieves the expected original sequence.

Peculiarly, we can also notice in Fig. 2(a) that, after 2 million episodes, the relative PM curiously decreases. Likewise, such behavior can be observed in several other experiments, as shown in Fig. 2(b), which presents the relative PM for all experiments performed on samples composed of $8bp$-long reads.

Through a more in-depth analysis of the control experiment demonstrated by Fig. 3, we can conclude that this behavior may be explained by the exploration-exploitation tradeoff, since in the initial episodes the chances of the agent taking random actions are greater. Figure 3 shows obtained PMs in any of the 3 runs of the control experiment, from the first training episode until 5 million episodes. In this execution, besides observing the stabilization of PM after 1 million episodes, it is possible to identify the high oscillation presented in the initial episodes. As most experiments have higher number of reads than the aforementioned experiment, the size of state space of these experiments is consequently larger, as we can see in the last row of Table 2, which presents the total number of states considering 10, 20, and 30 reads according to the investigated approach.

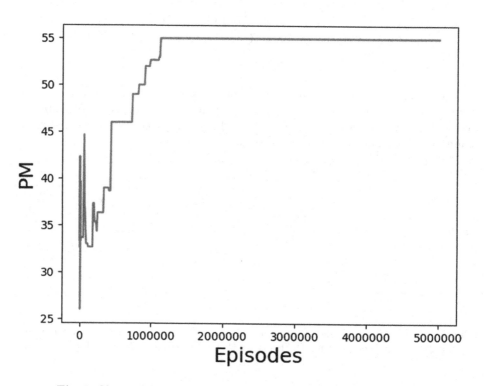

Fig. 3. Obtained PM across all episodes for the control experiment.

Consequently, the number of episodes required to reach the optimal convergence for 20 and 30 reads is probably much higher than the number of episodes required for the experiments with 10 reads, as we can see by the average number of states explored in experiments with 10, 20 and 30 reads in Table 2.

Since optimal convergence was achieved after 4 million episodes for the control experiment, we can see from Table 2 (in bold) that convergence was achieved after the exploration of approximately 0.02% of state space. Table 2 also presents (at the bottom) the expected number of states equivalent to 0.02% of the state space for 10, 20, and 30 reads. Note that this expectation is much higher than the actual number of states visited in the experiments after 5 million episodes.

Table 2. Average number of states visited during training (the first column shows the number of episodes in millions of episodes)

Episodes	10	20	30
1	1.094.876	10.354.122	20.357.849
2	1.715.598	20.088.167	39.921.756
3	2.175.564	29.560.838	59.302.879
4	**2.561.952**	38.835.584	78.531.234
5	2.892.546	47.989.471	97.511.423

0,02% →	2.561.952	22e21	42e39
100% →	≈ 1e10	≈ 1e26	≈ 2e44

Analyzing the growth trend of the number of states visited as a function of the number of episodes, it is possible to observe that the number of states increases linearly (although at different coefficients), as we can see in Fig. 4, which shows the number of states visited for the control experiment. The same linear behavior can be observed in the training time of the model.

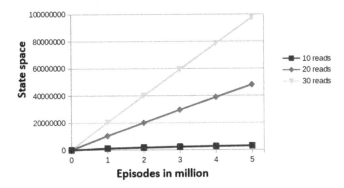

Fig. 4. State space exploration as a function of the number of episodes.

Assuming the exploration percentage of the first experiment as ideal to reach the optimal convergence and computing the linear regression for state-space exploration and time-consuming for 20 and 30 reads, we can estimate the number of episodes and time necessary to reach a percentage state corresponding.

Additionally, we can estimate the *Q-table* size to evaluate computational time and space needed.

According to these estimates, we can infer that $3.97e16$ episodes would be required to reach convergence in experiments with 20 reads—against $7.67e34$ episodes for 30 reads. The executions of these episodes, taking into account the respective regression coefficients of the observed training times, would require more than 180 billion years for the case of the 20 reads experiment[3].

From the perspective of required storage space, the estimates for optimal convergence are also not favorable, presenting computationally unreachable demands. When we consider the number of states presented in Table 2 as ideal to reach optimal convergence, even if we could store each value of the Q-table in a single bit, the space required to store the Q-table for 20 reads would be approximately 3 million petabytes if we assume the best case—in which only one action will be taken in each state visited; for 30 reads, it would be necessary about $50e30$ petabytes.

As mentioned earlier, the correct definition of rewards is critical to agent learning. In this sense, Bocicor et al. proposed the use of *PM* as a reward for actions that lead to transitions to final states. The metric was proposed assuming that maximizing *PM* implies optimal solutions in the datasets used.

However, such a premise is not always true, as we can observe in Table 1, where we can find some cases where the obtained *PM* was greater than the expected *PM*[4]. This situation can be better observed in Table 3, that presents the relative PM and the corresponding distance for a specific run. Note that, even presenting relative *PM* greater than 100%, the resulting genome requires, at least, 26 modifications to be transformed into the original microgenome.

Table 3. Relative *PM* and distance for the run performed on the sample with 10 reads of $8bp$ from the microgenome of $50bp$

Episode	Relative PM	Distance
1	104,98%	36
2	106,64%	26
3	105,01%	31
4	105,01%	31
5	106,64%	26

This is the reason why the viability analysis demonstrated here has been presented under these two perspectives so that only assemblies with high *PM* combined with low distance were considered good assemblies.

---

[3] Supplementary file 2 presents the training times of each experiment.

[4] Supplementary file 1 presents an example for better understanding such situation.

5   Conclusion

The development of a free assembler for RL-based genome assembly seems to be appropriate, although there are some aspects to be considered, among them the appropriate choice of the algorithm to obtain an optimal policy. As demonstrated, *Q-learning* requires a table whose size increases as the state space increases.

For real assembly scenarios—where the number of reads is immensely larger than the quantities used in the experiments presented here, we conclude there is no tangible time and storage space to properly train the agent using *Q-learning*, which becomes incapable of obtaining the correct order of sequences to reconstruct the original genome. Nevertheless, alternative training algorithms, especially those based on deep learning, have demonstrated great performance in tasks that require large state spaces.

A novel RL approach, termed a Deep Q-Network (DQN), to develop an intelligent agent that can successfully learn policies to play Atari 2600 games was introduced by Mnih et al. [6,7]. In this model, every state is represented as a combination of pixel values from four gray-scale frames, where each value varies between 0 and 255. It uses a convolutional network configured with an $84 \times 84 \times 4$ input layer, resulting in 28.224 neurons. Consequently, there are $256^{28.224}$ different possible combinations, equivalent to more than $1e67000$, which is a huge state space even compared with our most complex case, 30 reads, that contains less than $1e50$ possible states.

Reward is another aspect to be considered in modeling an RL-based model for assembly. We demonstrated that the maximization of the current metric used, PM, can lead the agent to wrong answers. Given that optimal policy learning is based on maximizing accumulated rewards, an in-depth analysis of the adequacy of this metric becomes reasonable.

References

1. Bellman, R.E., Dreyfus, S.E.: Applied Dynamic Programming, vol. 2050. Princeton University Press, Princeton (2015)
2. Bocicor, M.I., Czibula, G., Czibula, I.G.: A reinforcement learning approach for solving the fragment assembly problem. In: 2011 13th International Symposium on Symbolic and Numeric Algorithms for Scientific Computing. IEEE, September 2011
3. Heather, J.M., Chain, B.: The sequence of sequencers: the history of sequencing DNA. Genomics **107**(1), 1–8 (2016)
4. Li, Z., et al.: Comparison of the two major classes of assembly algorithms: overlap-layout-consensus and de-bruijn-graph. Briefings Funct. Genomics **11**(1), 25–37 (2011)
5. Miller, F.P., Vandome, A.F., McBrewster, J.: Levenshtein Distance: Information Theory, Computer Science, String (Computer Science), String Metric, Damerau? Levenshtein Distance, Spell Checker, Hamming Distance. Alpha Press (2009)
6. Mnih, V., et al.: Playing Atari with deep reinforcement learning. arXiv preprint arXiv:1312.5602 (2013)

7. Mnih, V., et al.: Human-level control through deep reinforcement learning. Nature **518**(7540), 529–533 (2015)
8. Pop, M.: Genome assembly reborn: recent computational challenges. Briefings Bioinform. **10**(4), 354–366 (2009)
9. Rangwala, H., Charuvaka, A., Rasheed, Z.: Machine learning approaches for metagenomics. In: Calders, T., Esposito, F., Hüllermeier, E., Meo, R. (eds.) ECML PKDD 2014. LNCS (LNAI), vol. 8726, pp. 512–515. Springer, Heidelberg (2014). https://doi.org/10.1007/978-3-662-44845-8_47
10. Shang, J., Zhu, F., Vongsangnak, W., Tang, Y., Zhang, W., Shen, B.: Evaluation and comparison of multiple aligners for next-generation sequencing data analysis. BioMed Res. Int. **2014**, 1–16 (2014)
11. Smith, T., Waterman, M.: Identification of common molecular subsequences. J. Mol. Biol. **147**(1), 195–197 (1981)
12. Soueidan, H., Nikolski, M.: Machine learning for metagenomics: methods and tools. arXiv preprint arXiv:1510.06621 (2015)
13. de Souza, K.P., et al.: Machine learning meets genome assembly. Briefings Bioinform. **20**(6), 2116–2129 (2018)
14. Sutton, R.S., Barto, A.G.: Reinforcement Learning: An Introduction. MIT Press, Cambridge (2018)
15. Watkins, C.J.C.H.: Learning from delayed rewards (1989)

GeNWeMME: A Network-Based Computational Method for Prioritizing Groups of Significant Related Genes in Cancer

Jorge Francisco Cutigi[1,3]($\boxtimes$), Adriane Feijo Evangelista[2], and Adenilso Simao[3]

[1] Federal Institute of Sao Paulo, Sao Carlos, SP, Brazil
`cutigi@ifsp.edu.br, cutigi@usp.br`
[2] Barretos Cancer Hospital, Barretos, SP, Brazil
`adriane.feijo@gmail.com`
[3] University of Sao Paulo, Sao Carlos, SP, Brazil
`adenilso@icmc.usp.br`

Abstract. Identifying significant mutations in cancer is a challenging problem in Cancer Genomics. Computational methods for identifying significant mutations have been developed in recent years. In this work, we present a flexible computational method named GeNWeMME (Gene Network + Weighted Mutations + Mutual Exclusivity). Our method uses an extensive biological base for prioritizing groups of significant and related genes in cancer. Our method considers data about mutations, type of mutations, gene interaction networks and mutual exclusivity pattern. All these aspects can be used according to the objective of the analysis by cancer genomics professionals, that can choose weights for each aspect. We test our method in four types of cancer where it was possible to identify known cancer genes and suggest others for further biological validation.

Keywords: Cancer bioinformatics · Driver mutations · Significant mutations · Computational method

1 Introduction

A cancer cell has a large number of somatic mutations. These mutations are, for the most of times, random and do not contribute to cancer, the so-called passenger mutations. While most somatic mutations are passenger mutations, there is a smaller group of mutations that are significant for cancer, the so-called driver mutations [17]. In this work, we consider the terms "significant mutations" and "driver mutations" as synonymous.

New genome sequencing technologies, called Next-Generation Sequencing (NGS), promote fast and cost-effective genomic sequencing. These new technologies enable the generation of a large volume of biological data in a short time.

© Springer Nature Switzerland AG 2020
L. Kowada and D. de Oliveira (Eds.): BSB 2019, LNBI 11347, pp. 29–40, 2020.
https://doi.org/10.1007/978-3-030-46417-2_3

These data are used by researchers to study and analyze genetic alterations in many diseases, including cancer. Many studies have resulted in the identification of a high number of recurrent mutations in cancer [20].

Studies using NGS data also have shown that a small number of genes are mutated with high frequency in a given set of patients and a high number of genes are low-frequency mutated [8]. Some mutated genes with a low frequency of mutation can be genes that are significant for cancer, which brings with it a statistical difficulty because it is not enough to mention the genes with the highest frequency of mutation as a driver mutation. This context shows that many significant genes have not yet been discovered since many of these genes appear at low-frequency [8]. One of the causes of this phenomena is the heterogeneity of cancer, which is the fact that two genomes of the same type of cancer do not necessarily have the same set of mutations.

Identifying significant genes for cancer is a big challenge on cancer genomics. In order to work in this challenge, computational methods have been developed to identify significant mutations in cancer. These methods have found new associated cancer genes from NGS data. The methods usually consider mutations that happened in a single gene (e.g., MutSig [11]) or in a group of them (e.g., MEMo [3]). The analysis of a group of genes is interesting due to cancer is the result of mutations in multiple genes.

In this work, we propose the GeNWeMME (Gene Network + Weighted Mutations + Mutual Exclusivity) method, which is a computational method for prioritizing group of related genes significantly mutated in cancer. We extend, modify, implement, and evaluate the work described in [4], which is a short paper that describes a proposal of a computational method that works with the manipulation of biological data and provides flexibility in the analysis. GeNWeMME considers three aspects: (1) Mutation data and the relevance of them; (2) The genes and the interaction among them; and (3) The mutual exclusivity pattern. In order to evaluate our method, we analyzed data of four types of cancer: glioblastoma, ovarian, prostate and lung cancer. In the results, we can observe that GeNWeMME is able to find important genes for cancer and to suggest others for new studies. Also, we show that GeNWeMME may prioritize genes with a low frequency of mutation.

The remaining of this paper is organized as follows. First, in Sect. 2, we present some related works. Second, in Sect. 3, we present the main biological concepts involved in this work. The proposed method is described in Sect. 4, where we present the steps, a running example and a complexity analysis. Next, in Sect. 5, we present the results of the method after applying two experiments. Finally, in Sect. 6, we present the final considerations about our work.

2   Related Work

The HotNet method [18] finds multiple sets of genes that are mutated in a relatively high number of patients. The algorithm creates a gene graph and uses the network diffusion algorithm to separate nodes (genes) and find relevant subsets.

As a criterion to find these subsets, it is defined how hot is a gene (how is your interaction in the network) and his coverage (how many times it was mutated in the patients). The HotNet2 method [13] extends the HotNet method. The diffusion process used in this method better encodes the network topology, where the method uses a directed graph to find significant subnetworks. The Hierarchical HotNet method [16] uses gene network and gene scores to construct a hierarchy topologically close and high-scoring subnetworks [16]. The MUFFIN method [2] generates a ranking of possible driver mutations, based on the gene networks and the mutation data from the genes and their neighbors in the network.

The Dendrix method [19] works with the hypothesis that pathways contain a set of genes that the mutation has a high coverage (most patients have at least one mutation in the set) and exhibits a high pattern of high exclusivity (most patients have only one mutation in the set). Dendrix method measures how much a set of genes has these characteristics defining a weight function to get this measure. Based on the measure, Dendrix method uses two approaches to select one set of genes that provide the best measure. The Multi-Dendrix method [12] uses the same weight function than Dendrix method, and it is capable of finding multiple driver pathways. The CoMEt method [14] treats a limitation of the Dendrix weight function, which happens when one or more genes have a high mutation frequency. To address this problem, the authors developed an exact statistical test for mutual exclusivity.

The MEMo method [3] performs statistical analysis and tests to identify characteristics in gene network modules based on three criteria. The objective of the MEMo method is to identify the set of related genes (modules) that are frequently altered, belonging to the same biological process with a pattern of mutual exclusivity. In this similar context, the MEMCover method [9] uses network analysis to identify sets of genes mutually exclusive in the same type of cancer and across many types of cancer.

Our method, GeNWeMME, could better fit in the same category of MEMo and MEMCover, although it uses ideas from Dendrix and HotNet. Moreover, the GeNWeMME method provides flexibility, where the user can adjust which biological aspects are more important for the analysis.

3 Biological Concepts

Gene interaction networks, or gene networks, represent complex interactions around genes and their produced proteins. In this kind of networks, genes are the nodes, and the edges connect the genes that are physically interacting or functionally related. Gene networks are largely used in cancer genomics, and there are many databases with gene interaction information.

Mutations in cancer occur in different scales, from a simple variation of a single nucleotide to a huge alteration in a significant part of the chromosome or even in the whole chromosome. SNVs (Single Nucleotide Variants) happens when a single nucleotide is substituted by the other. SNVs, can be *missense* or *nonsense*. Missense mutation results in the substitution of one amino acid for another. A nonsense mutation creates a signal to the cell stop building the protein, resulting in a shortened protein. InDels (Insert and Deletions) happens when a single or small sequence of nucleotides can be inserted or deleted of part of the DNA sequence. InDels cause *frameshift* mutations, which leads to errors in the reading sequence to produce the protein. *Splice* mutations are important mutations that occur in the regions where the splicing phase happens, in the creation of messenger RNA.

The mutual exclusivity pattern is related to a group of two or more genes, where the genes into this group are rarely mutated in the same patient, i.e., simultaneous mutations of these genes in the same patient are less frequent than they are expected by chance [10]. On the other hand, different genes of the group can be mutated in different patients.

4   GeNWeMME Method

In order to find a group of related genes significantly mutated in a cohort of patients, we propose GeNWeMME, a network-based method to prioritize groups of genes with size k. Our method has four steps and uses somatic mutation data (SNVs and InDels), gene interaction network and mutual exclusivity pattern. With these data and pattern, we find connected genes and we define a function to prioritize a group of related genes that are mutated in most patients with a high impact and present a pattern of mutual exclusivity. In Fig. 1 we present an overview of the GeNWeMME method.

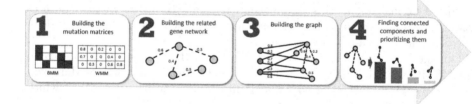

Fig. 1. An overview of the GeNWeMME method.

Step 1: Building the mutation matrices

In this first step, we filter genomic data in order to keep the most relevant genes in the analysis. For this, we choose a threshold ϕ, and we keep genes mutated in at least $\phi\%$ of the patients. After that, we build two matrices: (1) Weighted Mutation Matrix (WMM); and (2) Binary Mutation Matrix (BMM). In these matrices, the rows are patients and columns are genes.

In the WMM matrix wm, the entry wm_{ij} is assigned by a function $s(p_i, g_j)$ where, for each pair patient p_i and gene g_j, a score is obtained based on its type of mutation (Variant_Classification from MAF input file) and in a weight assigned for each type of mutation. Considering a patient p_i, and all mutations in a gene g_j, the score $s(p_i, g_j)$ is incremented by 0.5 if the mutation was Nonsense_Mutation, Frame_Shift_Del, or Frame_Shift_Ins. If the mutation is Missense_Mutation, Splice_Site, In_Frame_Ins, In_Frame_Del, 5'UTR, or Nonstop_Mutation, it is incremented 0.2. Finally, it is incremented by 0.1 if the mutation is Translation_Start_Site, or 3'UTR. After that, the final sum is divided by the number of mutations of the gene. For example, if the patient p_i has, in the gene g_j, three nonsense mutations, one frameshift mutation, three splice mutations, one missense mutation, and 2 in-frame mutation, the score $s(p_i, g_j) = ((3 + 1) \times 0.5 + (3 + 1 + 2) \times 0.2)/(3 + 1 + 3 + 1 + 2) = 0.32$. In summary, all pairs of a mutated gene g_j in a patient p_i has a score $s(p_i, g_j)$ that represents how important that mutation can be in that patient (although we define a weight for each type of mutation, the user can change them, through a input file). From the WMM, we can derive a BMM matrix bm, where the entry bm_{ij} has value 1 if wm_{ij} is greater than zero, and 0 otherwise. In the Fig. 2 we present a example of BMM and WMM. For the BMM, we represent the number 1 as a black cell and the number zero as a white cell. These matrices will be used on the running example in the next sections.

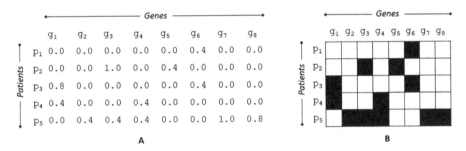

Fig. 2. (A) An Weighted Mutation Matrix (WMM); (B) A Binary Mutation Matrix (BMM).

Step 2: Building the related gene network

In this step, we build a Related Gene Network (RGN), that represents genes which are functionally related. Our method uses a gene interaction network from known databases to build the RGN. We initialize the RGN as the gene interaction network but placing the weight 1 in each edge. After that, for each gene pair $\{g_i, g_j\}$, we derive a measure using the Jaccard coefficient. For this measure, two genes g_i and g_j are considered proximal if they share a large number of common neighbors. Considering the set of neighbors of g_i and g_j as $N(g_i)$ and $N(g_j)$, respectively, we have the Jaccard coefficient calculated as follows: $J(g_i, g_j) = \frac{|N(g_i) \cap N(g_j)|}{|N(g_i) \cup N(g_j)|}$. We define a threshold γ, and if the measure $J(g_i, g_j)$ is greater than γ, then we include an edge between g_i and g_j, with the measure as its weight, i.e., we enrich the RGN with additional information based on the genes that shared a significant number of neighbors. This process is related to the approach presented by [3], where the authors use Jaccard coefficient to build a graph. In Fig. 3, we present the steps to produce the RGN. The produced RGN illustrated in the figure will be used on the running example in the next sections.

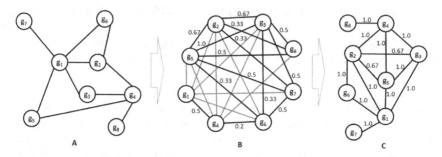

Fig. 3. (A) A hypothetical gene network, where nodes are genes and edges are inter-actions among them; (B) A Related Gene Network (RGN) build through applying the Jaccard coefficient in the original gene network. The gray edges represent the original interaction from the gene network (edge weight = 1), and the black edges are the new interactions based on the shared neighbors. (C) An RGN after to apply a threshold $\gamma = 0.6$, eliminating all the edges less than γ

Step 3: Building the graph

In this step, we build a graph $\mathcal{G}$, where the right side of the graph are the genes and their interactions (RGN), and the left side are the patients. To build $\mathcal{G}$, we use the wm and bm matrices obtained in Step 1, and the gene network RGN obtained in Step 2. In the bm, when a mutation is observed in a gene g_j in a specific patient p_i, an edge is included in $\mathcal{G}$ linking the patient p_i with the gene g_j. It is defined a weight in this created edge, where the weight is the corresponding entry wm_{ij}. In Fig. 4, we present an example of the graph $\mathcal{G}$, where it was used data from the matrices of the Fig. 2 and the RGN of the Fig. 3.

Step 4: Finding connected components and prioritizing them

After the Step 3, we have a graph $\mathcal{G}$ with important information contained in itself. Now, in the Step 4, we find a set cc of connected components of size k in the gene nodes of the graph. A k−connected component in a graph is a subgraph of size k, where there is at minimum one path between all the pair of the k nodes, i.e., all nodes of the subgraph are connected to each other by their paths. For example, in the Fig. 4, considering $k = 3$, the component $c_1 = \{g_4, g_5, g_8\}$ is connected, but $c_2 = \{g_1, g_3, g_8\}$ is not. Considering the RGN of the running example, we can extract 23 components of size $k = 3$.

Next, we prioritize the components found. The prioritization is based on three aspects, considering the set of the genes in one component c of the set cc. For this, we use three aspect functions $I(c)$, $R(c)$, and $W(c)$. For the next examples we will consider the component $c = \{g_1, g_3, g_5\}$:

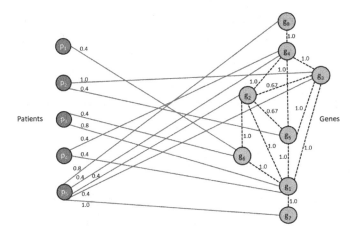

Fig. 4. An example of the graph $\mathcal{G}$. The dashed edges are interactions among genes, and the line edges are among genes and patients.

- The relevance of the mutation in the pair gene-patient: for this aspect, we define the function $I(c)$, that is the sum of the edge weights that link genes and patients. For the running example we have that $I(c) = 0.8 + 0.4 + 1.0 + 0.4 + 0.4 = 3.0$.
- How related are the genes: in this aspect, we define the function $R(c)$, that is the sum of the weights of the edges of c, considering only the gene nodes of the graph and their edges. For the running example we have that $R(c) = 1.0 + 1.0 + 1.0 = 3.0$.
- Mutual exclusivity pattern: to evaluate and quantify the mutual exclusivity among the genes in the component, we use the weight function $W(G)$ defined in the Dendrix method [19]. The function considers the trade-off between the coverage and exclusivity of a set of genes, looking for mutual exclusivity in the BMM. The weight function is defined by $W(G) = 2|\Gamma(G)| - \sum_{g \in G} |\Gamma(g)|$, where $\Gamma(g)$ are the patients where g is mutated and $\Gamma(G)$ are the patients where at least one gene in G is mutated. For the component c, we have $\Gamma(\{g_1, g_3, g_5\}) = 4$, $\Gamma(g_1) = 2$, $\Gamma(g_3) = 2$, and $\Gamma(g_5) = 1$, resulting in a $W(c) = 2 \times 4 - (2 + 2 + 1) = 3$.

We apply these aspect functions for all components of cc, and then we normalized all results by the max value of the elements, in order to keep all the values between 0 and 1 and improve the data presentation. Thus, we called the normalized result of each aspect function as $I'(c)$, $R'(c)$, and $W'(c)$. Next, we define a rank function $F(c)$, which considers all the aspects presented, i.e., the rank function combines the result of these three normalized aspect functions. For each aspect function, we have to choose a parameter in order to define the weights of each aspect. In this way, the rank function $F(c)$ is defined as $F(c) = \alpha I'(c) + \beta R'(c) + \delta W'(c)$. With this, we can define the parameters α, β, and δ according to the interest of the investigation. For example, if we are more interested in considering the mutual exclusivity pattern, it is expected to define the δ parameter greater than others. This makes the method flexible. Finally, for each component c in the set of components cc, the rank function $F(c)$ is applied. Then, according to the result of the function, the set of the genes is sorted in decreasing order. In Table 1, we present the final result for the running example, considering the five components with higher scores. It was defined the value 1 for the parameters α, β, and δ, i.e., the same weight for all parameters.

Table 1. The five components with higher score considering the running example and the parameters α, β, and δ as value 1.

Component	$I(c)$	$I'(c)$	$R(c)$	$R'(c)$	$W(c)$	$W'(c)$	$F(c)$
$\{g_1, g_3, g_6\}$	3.40	0.94	2.00	0.67	4	1.00	2.61
$\{g_1, g_3, g_5\}$	3.00	0.83	3.00	1.00	3	0.75	2.58
$\{g_1, g_2, g_3\}$	3.00	0.83	2.67	0.89	3	0.75	2.47
$\{g_1, g_2, g_5\}$	2.00	0.55	2.67	0.89	4	1.00	2.44
$\{g_1, g_2, g_6\}$	2.40	0.67	3.00	1.00	3	0.75	2.42

Complexity Analysis. The most representative operation in the proposed method is to find the connected components of size K in the RGN. Thus, in this complexity study, we will focus on this operation. We consider the parameter N as the number of genes in the RGN and K the size of the component that we are interested. Also, we define the parameter M as the average of edges of each gene in the RGN.

To build a component of size K, we start from one single gene. From this gene, we have to choose the second gene, and then we have M possibilities to choose. After the choice, we have more M possibilities, until to reach $K - 1$ choices. As a result, for one single gene, we have M^{K-1} numbers of components. As we have N genes in the RGN, we have, in total, $N \times M^{K-1}$ components to find. So, in big O notation, we have the complexity $O(NM^K)$.

We can notice that the problem has an exponential growth, which may be considered a limitation of the method. For our experiments, we did not have problems, since we chose a small K and the pre-processing routines resulted in a reasonable number of genes. Otherwise, we should think in heuristic approaches to find a solution in polynomial time, for any N, M and K.

5   Results

To evaluate our method, we used real genomic data about mutations and gene networks. We analyzed our results in comparison with known benchmarks. Although there is no gold standard for driver genes, some databases are widely used and continuously updated. As benchmark, we adopted a list of 739 known cancer genes from the Network of Cancer Genes (NCG6.0) [15] and from Cancer

Gene Census (CGC) [7]. As gene network we adopted the data from ReactomeFI (Version 2018) [6], in which we consider gene interactions with a score greater than 0.90, resulting in a network with 11,779 genes and 222,343 interactions.

For the analysis, we use somatic mutation data from four types of cancer. All data sets were extracted from other studies. For glioblastoma (GBM) we analyzed 145 patients with 1064 somatic mutations in 429 genes [3]. Considering ovarian cancer, we analyzed 324 patients with 15054 somatic mutations in 8424 genes [3]. For prostate cancer analysis, we analyzed data with 112 patients with 4185 somatic mutations in 3349 genes [1]. For the lung cancer data, we had 163 patients with 999 somatic mutations in 353 genes [5].

Experiment 1. In this experiment, we analyzed the precision of our method, that is obtained by $Precision = \frac{TP}{TP+FP}$. The variable TP are the number of true positives, i.e., the number of genes prioritized by our method that are in the benchmark. The variable FP are the number of false positives, i.e., the number of genes prioritized that are not in the benchmark. For each type of cancer, we obtained the precision considering as result the top N genes that appear in the group of genes. For example, considering the results of the running example, the top 4 genes are g_1, g_3, g_6, and g_5.

As the parameters of GeNWeMME, for the Step 1, we used $\phi = 2\%$. In Step 2, we used $\gamma = 0.05$ as the threshold for the RGN edges. Finally, in the Step 4 we defined components of size $K = 5$ and adopted the same values for the parameters α, β, and δ. To evaluate and compare the results, for each set N of genes, we generate 100 random sets, also checking if they are in the benchmark. The random set is composed by genes that are mutated in each type of cancer and that are in the network. We present the results in the Fig. 5. We can notice that, unsurprisingly, the GeNWeMME outperforms the random choice of genes. It is important to state that 100% of precision is not a desirable result, once we are interested in looking for new driver genes, and the benchmark we use to compare contains only known drivers. In this way, the GeNWeMME method is able to suggest new potential candidates for further studies. In Table 2, we show the top 10 prioritized genes, where the bold names are genes that are in the benchmark.

Table 2. The top 10 genes prioritized by GeNWeMME.

Cancer	Prioritized genes	Ratio
GBM	**TP53 PTEN PIK3R1 PIK3CA EGFR ERBB2 NF1** PIK3CG PSMD13 **PDGFRA**	8/10
Ovarian	**BRCA1 TP53 CREBBP RB1** TOP2A **BRCA2 NF1** PRKDC **CHD4 CDK12**	8/10
Prostate	**TP53 PTEN** TAF1L **SPOP CDKN1B FOXA1 PIK3CA ATM** VCAN ATP1A4	7/10
Lung	**ERBB4 NF1 KRAS** PIK3CG **EGFR** PIK3C3 PRKCG **JAK2 FGFR4 TP53**	7/10

Experiment 2. A challenge on identifying significant mutations in cancer is to prioritize genes with a low frequency of mutation. In this experiment, we evaluated the ability of the GeNWeMME method on prioritizing low-frequency genes. In the graphs of the Fig. 6 are presented the frequency of mutation from the top 10 genes prioritized by our method. In comparison, we show the distribution of the frequency of all genes of the input data. We can notice that not all genes with high frequency are prioritized by our method, while other genes with low frequency are. We used the same parameters of the first experiment.

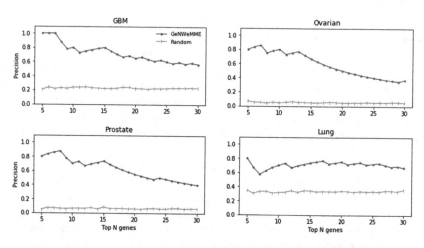

Fig. 5. Precision of the GeNWeMME method for finding known cancer drivers

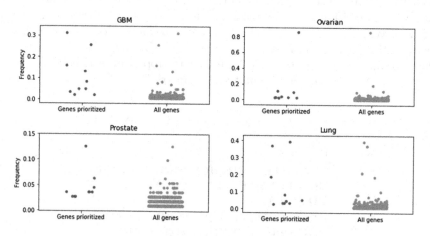

Fig. 6. The frequency of mutation of the top 10 genes prioritized by GeNWeMME.

6   Conclusion

This paper describes the GeNWeMME method, a computational method for prioritizing significant group of related genes in cancer. The method uses mutation data and gene network, also the evaluation of the mutual exclusivity pattern. Our contribution is to put these data together and consider all of them in the identification of possible significant set of genes in cancer. The method has parameters that can be adjusted according to the goal of the analysis. We provide this flexibility to allow professionals from cancer genomics are able to use the method according to the characteristics of their experiments.

We show that the GeNWeMMe method is able to prioritize genes that are already known to be driver, while the method suggests other genes for profound studies and biological experiments. We also show that the method is able to prioritize genes with very low mutation frequency. The results were obtained on applying GeNWeMME in four types of cancer.

As future work, the integration of more biological data can be a promising idea, for example, gene expression, and mutational signatures. A more systematic and extensive experimental evaluation is needed, for example, to make a comparison with related methods, performing experiments with other types of cancer, and allowing the use of copy number mutation data. Furthermore, a functional enrichment evaluation of the recovered genes may be performed, in order to discuss the biology plausibility of this result.

Source code, examples, cancer data, and the scripts of the experiments are available on the following link: https://github.com/jcutigi/GeNWeMME_BSB2019.

References

1. Barbieri, C.E., et al.: Exome sequencing identifies recurrent SPOP, FOXA1 and MED12 mutations in prostate cancer. Nature genetics **44**(6), 685–689 (2012). https://doi.org/10.1038/ng.2279
2. Cho, A., Shim, J.E., Kim, E., Supek, F., Lehner, B., Lee, I.: MUFFINN: cancer gene discovery via network analysis of somatic mutation data. Genome Biol. **17**(1), 129 (2016). https://doi.org/10.1186/s13059-016-0989-x
3. Ciriello, G., Cerami, E., Sander, C., Schultz, N.: Mutual exclusivity analysis identifies oncogenic network modules. Genome Res. **22**, 398–406 (2012)
4. Cutigi, J.F., Evangelista, A.F., Simao, A.: A proposal of a graph-based computational method for ranking significant set of related genes in cancer. In: Anais do XIX Simposio Brasileiro de Computacao Aplicada a Saúde, pp. 300–305. SBC, Porto Alegre, RS, Brasil (2019)
5. Ding, L., et al.: Somatic mutations affect key pathways in lung adenocarcinoma. Nature **455**(7216), 1069–1075 (2008)
6. Fabregat, A., et al.: The reactome pathway knowledgebase. Nucleic Acids Res. **46**(D1), D649–D655 (2018). https://doi.org/10.1093/nar/gkx1132
7. Futreal, P.A., et al.: A census of human cancer genes. Nat. Rev. Cancer **4**(3), 177–183 (2004)

8. Garraway, L.A., Lander, E.S.: Lessons from the cancer genome. Cell **153**(1), 17–37 (2013). https://doi.org/10.1016/j.cell.2013.03.002

9. Kim, Y.A., Cho, D.Y., Dao, P., Przytycka, T.M.: MEMcover: integrated analysis of mutual exclusivity and functional network reveals dysregulated pathways across multiple cancer types. Bioinformatics **31**(12), i284–i292 (2015). https://doi.org/10.1093/bioinformatics/btv247

10. Kim, Y.A., Madan, S., Przytycka, T.M.: WeSME: uncovering mutual exclusivity of cancer drivers and beyond. Bioinformatics **33**(6), 814–821 (2017). https://doi.org/10.1093/bioinformatics/btw242

11. Lawrence, M.S., et al.: Mutational heterogeneity in cancer and the search for new cancer-associated genes. Nature **499**, 214–218 (2013)

12. Leiserson, M.D.M., Blokh, D., Sharan, R., Raphael, B.J.: Simultaneous identification of multiple driver pathways in cancer. PLOS Comput. Biol. **9**(5), 1–15 (2013). https://doi.org/10.1371/journal.pcbi.1003054

13. Leiserson, M.D.M., et al.: Pan-cancer network analysis identifies combinations of rare somatic mutations across pathways and protein complexes. Nat. Genet. **47**(2), 106–114 (2015). https://doi.org/10.1038/ng.3168

14. Leiserson, M.D., Wu, H.T., Vandin, F., Raphael, B.J.: CoMEt: a statistical approach to identify combinations of mutually exclusive alterations in cancer. Genome Biol. **16**(1), 160 (2015). https://doi.org/10.1186/s13059-015-0700-7

15. Repana, D., et al.: The network of cancer genes (NCG): a comprehensive catalogue of known and candidate cancer genes from cancer sequencing screens. Genome Biol. **20**(1), 1 (2019). https://doi.org/10.1186/s13059-018-1612-0

16. Reyna, M.A., Leiserson, M.D.M., Raphael, B.J.: Hierarchical hotnet: identifying hierarchies of altered subnetworks. Bioinformatics **34**(17), i972–i980 (2018). https://doi.org/10.1093/bioinformatics/bty613

17. Stratton, M.R.: The cancer genome. Nature **458**(7239), 719–724 (2009). https://doi.org/10.1038/nature07943

18. Vandin, F., Upfal, E., Raphael, B.J.: Algorithms for detecting significantly mutated pathways in cancer. J. Comput. Biol. **18**(3), 507–522 (2011). https://doi.org/10.1089/cmb.2010.0265

19. Vandin, F., Upfal, E., Raphael, B.J.: De novo discovery of mutated driver pathways in cancer. Genome Res. **22**(2), 375–385 (2012). https://doi.org/10.1101/gr.120477.111

20. Vogelstein, B., Papadopoulos, N., Velculescu, V.E., Zhou, S., Diaz, L.A., Kinzler, K.W.: Cancer genome landscapes. Science **339**(6127), 1546–1558 (2013). https://doi.org/10.1126/science.1235122

MDR SurFlexDock: A Semi-automatic Webserver for Discrete Receptor-Ensemble Docking

João Luiz de Almeida Filho$^{(\boxtimes)}$ (iD) and Jorge Hernandez Fernandez (iD)

LQFPP, Center of Biosciences and Biotechnology,
State University of North Fluminense, Campos dos Goytacazes, RJ, Brazil
joaoluiz.af@gmail.com, jorgehf@uenf.br

Abstract. In current computational biology, docking is a popular tool used to find the best fit of one ligand relative to its molecular receptor in forming a complex. However, the most of the tools do not take into account the flexibility of the receptor due to computational cost. As a result, the conformational changes caused by the induced fit are ignored in exploratory docking experiments. In this context and to improve the predictive capacity of docking, a good strategy is to simulate the flexibility of the receptor with the use of key conformations that can be obtained by mixing crystallography and computer simulations, a technique known as ensemble docking. Here, we present MDR SurFlexDock, a web tool that improves the docking experiments by computing a discrete, but representative, ensemble of contact surfaces of the receptor through clustering of molecular simulation trajectories in order to simulate the intrinsic flexibility of the ligand-contacting surface. The results of the interaction of each receptor-compound complex are presented in a concise tabular format to allow rapid analysis of compounds when classifying them by inhibition constant (Ki). MDR SurFlexDock can be valuable in cases of docking for new receptors obtained by homology modelling, in extensive analysis of different chemotypes on proteins with low structural information and for fast characterization of binding capacities on contact surfaces with poor structural information or only optimized for a specific ligand. MDR SurFlexDock is freely available as a web service at http://biocomp.uenf.br:81.

Keywords: Drug design · Ensemble docking · AutoDock

1 Background

Docking is one of most popular techniques in computer-aided drug design [1,2]. This approach can be summarized as the discovery of the compounds (or ligands) that best fit into a specific part of the target protein (the receptor) [3]. As part of the experiment, structural information and interaction energy of the complex are important factors regarding the final choice for the 'best interaction model',

© Springer Nature Switzerland AG 2020
L. Kowada and D. de Oliveira (Eds.): BSB 2019, LNBI 11347, pp. 41–47, 2020.
https://doi.org/10.1007/978-3-030-46417-2_4

as most of the docking programs provide several mathematical solutions for the same problem [2], and the best results, in most cases, is the final user choice.

As proteins are flexible and dynamic macromolecules that continually alternate their surface conformations, the accurate representation of the protein surface facing the ligands is currently one of the most important challenges in docking experiments [4]. At the same time, simulating the receptor flexibility increases the computational cost of the experiment. To overcome the problem of protein flexibility several strategies are used to better sample the conformational space of the receptor [2]. Moreover, the main protein structural information of the receptor contact surface is predefined by complexes that are resolved by X-ray crystallography and stored in databases (PDB format at www.rcsb.org). As a result, protein surfaces in complexes with some ligand are typically biased to perform better in some specific contacts, and docking experiments with different chemotypes becomes a challenging task [4].

In this context, we are developing MDR SurFlexDock (Molecular Dynamics-based Receptor Surface Flexibility for Docking), a web server that allows analysis of the interaction of small protein-ligand complexes through enhanced structural sampling of receptor surfaces with low computational costs. The pipeline uses of discrete conformational search in solvated receptors, performing molecular simulations after gradual thermalization and increasing the possibilities to better represent the physiological environment in several protein-ligand docking experiments. The receptor surface is submitted to clustering analysis over the simulation trajectories with a 1.0 Å cut-off. Finally, the three most representative conformations are used in consecutive docking experiments. At the end of the experiment, MDR SurFlexDock presents the results through a concise table of protein-ligand complexes, which are classified by the inhibition constant (Ki), as well as discrete graphical information on receptor clustering and statistical representation of ten better dockings for each compound.

2   Implementation

2.1   Web Server

The main page of MDR SurFlexDock (Fig. 1:A) provides a simple and convenient way to specify the target receptor in pdb format, up to ten ligands in the native AutoDock format (.pbdqt), an email for user notifications and docking parameters defining receptor contacting surface (Fig. 1:B). The server will redirect the user to an experiment control page (Fig. 1:C) and, depending on the stage of experimentation, the control page will be fed on graphical information of the clustering, a box-plot plot of the top ten poses of each ligand in each of the docking experiments and a list of the calculated poses by MDR SurFlexDock (Fig. 1:C). In addition, structural information of each complex can be reached through a Glmol HTML 5 plugin [5]. Finally, MDR SurFlexDock provides a link to download all the experimental results to enable the user to handle them locally. Experimental results remain available for local download for over 15 days. The main page also provides a brief user-manual (http://biocomp.uenf.br:81/instructions) and a representative study case (http://biocomp.uenf.br:81/study_case) for interested users.

The MDR SurFlexDock server was developed using the Django web framework and BioPython plug-in. When a user starts an experiment, our server places experimentation in a queue managed by the Django-celery plug-in, redirects the user to an experiment specific page and e-mails the link to this page. In addition, Django is compatible with several Python libraries that were used in the development of this web server.

2.2   Pipeline

Due to the magnitude of the conformational space in proteins, taking into account the flexibility of the receptor in docking experiments constitutes a great challenge [2]. This becomes especially complex in the case of high-throughput screening (HTS). The docking of hundreds of compounds in a flexible receptor requires, even today, computational power not always available for newly interested users. One strategy is to use some conformations of the receptor to do the docking, which is known as ensemble docking [6]. MDR SurFlexDock uses a simple approach to ensemble docking in a two-stage pipeline: (i) initially constructing an ensemble containing the three most representative conformations of the receptor active site on the trajectory of 5-ns molecular simulation using the GROMACS 4.6 [7] with a GROMOS96 53a6 force field [8], followed by (ii) molecular docking of each compound with the ensemble of representative receptor structures using AutoDock 4.2 [9] and ADT 1.6 scripts [10] (Fig. 1).

Molecular simulation starts with the solvation of the receptor with a SPC216 water model [11] in a triclinic box, neutralized by randomly added Cl- and Na+ ions. The system energy is minimized in 50,000 steps, or reach up to 1,000 KJ/mol/nm, using the steep descent (SD) algorithm [12]. Receptor thermalization is done by heating the system temperature from 285 K to 300 K in the NVT ensemble. Subsequently, a 5-ns molecular simulation optimizes the interaction of the solvent with the receptor, thus exploring the conformational space of the receptor surface. Receptor samples are obtained by clustering the conformations of the user-defined active site using 0.1 nm RMSD cut-off and Gromos algorithm [13]. In this step, the active site is composed of all the residues that are at a certain (user-defined) distance in nanometres from a central residue, also defined by the user (Fig. 1:B).

The central structures of the three most representative clusters over the simulation trajectory are converted into pdb format to compose the ensemble (Fig. 2). As a control, docking of the original receptor structure delivered by the user is also performed with the ligands.

The docking procedure starts with the conversion of the receptor file into the native AutoDock format (.pdbqt) through ADT's 'prepare_receptor.py' script. Subsequently, each compound is verified by the script 'prepare_ligand4.py.' The Grid map is prepared through 'prepare_gpf4.py,' and the grid box is defined as a cube containing the residues of the active site that are a certain distance from the central residue (in nanometres). Both parameters are pre-defined by the user.

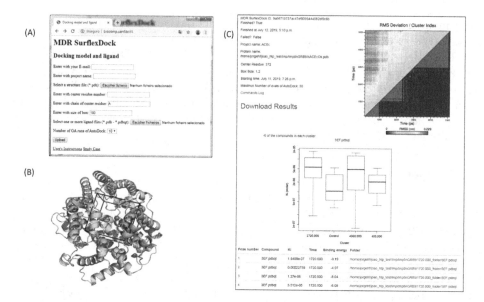

Fig. 1. Screenshots of MDR SurFlexDock. (A) Main page. (B) The active site is defined by the user as cubic box using a central residue and box size. (C) Results page containing information about the catalytic site clustering, inhibition constant and poses found. In HTP SurflexDock the results page is fed during the experiment.

This script generates the file 'grid.gpf,' which is executed by the autogrid4.2 software [9]. Subsequently, the docking experiment is configured by the script 'prepare_dpf4.py' to create the amount of poses defined by the user (10, 20 or 30) and 2.5 x 10E6 energy inferences for every docking experiment, using the Lamarckian genetic algorithm (LGA) [9]. Finally, the poses are calculated using the autodock4.2 software. During each docking, the docking log-file (.dlg) is parsed by the BioPython library to extract the inhibition constant (Ki) that will be used in the experimental output (Fig. 1:C). The MDR SurFlexDock pipeline is freely available as a web service at http://biocomp.uenf.br:81. A general flowchart of the experimental pipeline is represented in Fig. 2.

3   Results and Discussion

Here we describe MDR SurFlexDock, a virtual screening tool that addresses the problem of receptor molecular sampling through ensemble docking. One of the main goals of MDR SurFlexDock is to be a simple tool to use for exploring the initial stages of protein-ligand studies, at which point no experimental and structural information is available, and the need for an initial hypothesis of a protein-ligand structural interaction is mandatory. When implemented, even for homology modelling of the receptor, minimal information about the active site and a set of ligands in .pdbqt format is sufficient for starting experiments. In its

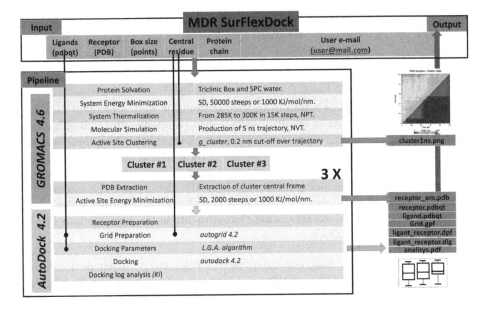

Fig. 2. MDR SurFlexDock pipeline. The user inputs a receptor protein in pdb format, with one residue representing the centre of the active site, and up to ten ligands of interest in pdbqt format. The molecular simulations and clustering over trajectory is executed by GROMACS 4.6, and the docking calculations are performed using AutoDock 4.2. Each experiment has its own experimental workspace containing information about the submission, progress and a listing of the calculated poses classified by Ki constant.

actual developmental stage, MDR SurFlexDock allows dockings with up to ten ligands per experiment due to limited hardware resources. The construction of extensive ensembles using molecular dynamics may be an important limitation of this approach, due to the high computational cost [6]; however, some studies indicate that positive results can be achieved if the simulation is limited to some residues [14]. In the MDR SurFlexDock pipeline, this process is performed by clustering only the user-defined contact surface under rapid molecular simulation across the receptor.

In this regard, our tool constructs the most representative ensemble, with discrete structural distances between the conformations and specific differences in the ligand-interacting surface, although it does not identify large movements (such as hinged effects) due to limitations in molecular simulation time. We aimed to maintain a low computational cost, since as the number of ensemble conformations increases, the amount of dockings and experimental time will increases at the ratio of geometric progression. In addition, a greater number of ensemble conformations in a set increase the likelihood of an anti-cooperative effect [6]. In the case of MDR SurFlexDock, this would result in a waste of computer time due to docking calculations for these conformations and an unnecessary increase in complexity of experimental results.

4 Perspectives

Our group herein describes a simple and semiautomatic web implementation for analysis of the interaction of small protein-ligand complexes through enhanced structural sampling of the receptor surface. In principle, this technique can be used in MDR experiments using one of the several ligand databases available (http://autodock.scripps.edu/resources/databases), increasing the total number of positive results and reducing false negatives. We also foresee the integration of a better parameterization tool for ligands and the possible implementation of alternative scoring functions for docking experiments in the pipeline in a future release of MDR SurFlexDock.

Funding. This work has been supported by the Conselho Nacional de Desenvolvimento Científico e Tecnológico (CNPq) doctoral grant [141917/2015-6] for J.L.A.F. and ProAP-CAPES support from the Coordenação de Aperfeiçoamento de Pessoal de Nível Superior (CAPES), which are gratefully acknowledged.

References

1. Trott, O., Olson, A.J.: AutoDock Vina: improving the speed and accuracy of docking with a new scoring function, efficient optimization, and multithreading. J. Comput. Chem. **31**, 455–461 (2010)
2. Pagadala, N.S., Syed, K., Tuszynski, J.: Software for molecular docking: a review. Biophys. Rev. **9**(2), 91–102 (2017). https://doi.org/10.1007/s12551-016-0247-1
3. Ouzounis, C.A.: Rise and demise of bioinformatics? Promise and progress. PLoS Comput Biol. **8**, e1002487 (2012)
4. Guedes, I.A., de Magalhães, C.S., Dardenne, L.E.: Receptor-ligand molecular docking. Biophys. Rev. **6**, 75–87 (2014)
5. Nakane, T.: GLmol-Molecular Viewer on WebGL/Javascript, Version 0.47 (2014)
6. Antunes, D.A., Devaurs, D., Kavraki, L.E.: Understanding the challenges of protein flexibility in drug design. Expert Opin Drug Discov. **10**, 1301–1313 (2015). https://doi.org/10.1517/17460441.2015.1094458
7. Abraham, M.J., Murtola, T., Schulz, R., Páll, S., Smith, J.C., Hess, B., et al.: GROMACS: high performance molecular simulations through multi-level parallelism from laptops to supercomputers. SoftwareX **1**, 19–25 (2015)
8. Oostenbrink, C., Soares, T.A., der Vegt, N.F.A., Van Gunsteren, W.F.: Validation of the 53A6 GROMOS force field. Eur. Biophys. J. **34**, 273–284 (2005)
9. Norgan, A.P., Coffman, P.K., Kocher, J.-P.A., Katzmann, D.J., Sosa, C.P.: Multi-level parallelization of AutoDock 4.2. J. Cheminform. **3**, 12 (2011)
10. Morris, G.M., et al.: AutoDock4 and AutoDockTools4: automated docking with selective receptor flexibility. J. Comput. Chem. **30**, 2785–2791 (2009)
11. Teleman, O., Jönsson, B., Engström, S.: A molecular dynamics simulation of a water model with intramolecular degrees of freedom. Mol. Phys. **60**, 193–203 (1987)
12. Fletcher, R.: Practical Methods of Optimization. Wiley, Hoboken (2013)

13. Daura, X., Gademann, K., Jaun, B., Seebach, D., Van Gunsteren, W.F., Mark, A.E.: Pep-tide folding: when simulation meets experiment. Angew. Chem. Int. Ed. **38**, 236–240 (1999)
14. Armen, R.S., Chen, J., Brooks III, C.L.: An evaluation of explicit receptor flexibility in molecular docking using molecular dynamics and torsion angle molecular dynamics. J. Chem. Theory Comput. **5**, 2909–2923 (2009)

Venom Gland Peptides of Arthropods from the Brazilian Cerrado Biome Unveiled by Transcriptome Analysis

Giovanni M. Guidini[1], Waldeyr M. C. da Silva[1,2](✉) ⓘ, Thalita S. Camargos[1],
Caroline F. B. Mourão[1], Priscilla Galante[1], Tainá Raiol[3],
Marcelo M. Brígido[1]ⓘ, Maria Emília M. T. Walter[1]ⓘ,
and Elisabeth N. F. Schwartz[1]

[1] University of Brasília (UnB), Brasília, DF, Brazil
giovanni.guidini@gmail.com, thalitasoares@gmail.com,
caroline.barbosa@ifb.edu.br, prigalante@yahoo.com.br,
{brigido,mariaemilia}@unb.br, beth.ferroni@gmail.com
[2] Federal Institute of Goiás, Formosa, GO, Brazil
waldeyr.mendes@ifg.edu.br
[3] Fiocruz, Brasília, DF, Brazil
tainaraiol@gmail.com

Abstract. Animal venoms are rich sources of pharmacological active molecules. Less than 10% of arthropod venom components have been characterized so far, reinforcing the importance of prospective studies. The Cerrado, in the Midwest Region of Brazil, is the second-largest biome in Brazil presenting vast biodiversity of arthropod species with venom glands. In this scenario, in a project called "Inovatoxin", active principles present in the venom of three biodiversity representative arthropod animals from this region were characterized structurally and functionally, using proteomic and transcriptomic prospective strategies. High Throughput Sequencing (HTS) is among the strategies to provide the raw material to help identify bioactive peptides present in these arthropods' venom. This work proposes a workflow that allowed to annotate a total of 230 venom peptides from the Brazilian arthropods spider *Acanthoscurria paulensis*, social wasp *Polybia sp.*, and scorpion *Tityus fasciolatus*. Along with these results, abundant data on the metabolism of the three species were also obtained. These results extend knowledge of venoms, contributing to new perspectives on rational therapeutic measures to treat accidents with these animals, and also on academic and biotechnological applications.

Keywords: Animal venom · Toxin · Scorpion · Social wasp · Spider · Arthropods · Biodiversity · Cerrado biome

---

G. M. Guidini and W. M. C. Silva—contributed equally to this work.

L. Kowada and D. de Oliveira (Eds.): BSB 2019, LNBI 11347, pp. 48–57, 2020.
https://doi.org/10.1007/978-3-030-46417-2_5

1   Introduction

Animal venoms have been recognized as significant assets of biologically active molecules. It is estimated that the *Arthropoda* is the most diverse animal group on the planet [27]. Although Brazilian biodiversity covers approximately 20% of the total species of the planet [11], much of this biotechnological potential remains unknown, notably in Brazilian Cerrado, which is the second-largest Brazilian biome [25,28].

Cerrado biome is the major habitat type of Brazilian Midwest Region, including the states of Mato Grosso (MT), Mato Grosso do Sul (MS), Goiás (GO), and the Federal District (DF). In this region, *Tityus fasciolatus* is an endemic scorpion species and it is implicated in most of the reported scorpion accidents [20].

Also, social wasps as *Polybia sp.* are an important, diversified and an abundant component of the Brazilian fauna with social behavior, presenting an architectonic diversity of their nests [19,25]. Most of the studies developed in the Cerrado areas include social wasps faunistic composition and abundance, population density, nesting habitats, and seasonality [25]. Thus, the antinociceptive activity of insect venom, including *Polybia sp.*, is a promising area [16].

Besides, venom extracted from the Brazilian spider *Acanthoscurria paulensis* has demonstrated pharmacological activity [17]. There are more than 2 million polypeptides in the venom of arthropod species, and most of them remain to be characterized [27].

The biotechnological potential of animal venoms includes proposing new pharmacological products through the study of toxins. Proteomic and transcriptomic are distinguished combined approaches to deeply understand and describe venom composition and the biological context of venoms for neglected species, and explore this potential [1,21,22].

Inovatoxin [24] is a project that aims to expand the knowledge of the venomous arthropod fauna of the Brazilian Cerrado Biome. The project was divided into two main paths: (i) development of proteomic analysis methodologies; combined with (ii) analysis of venom gland transcripts. This combination of strategies allows reaching an extensive outline of pharmacological and therapeutic potential of these venoms with a reduced number of specimens of each studied species, which favors sustainable exploitation of the arthropod fauna of the Brazilian Cerrado. Withal, these studies represent crucial contributions to improve our understanding of venom diversity on a broader scale.

Accordingly, this work describes a specially designed workflow, together with its results on toxins annotated from the venom gland transcripts of three species of arthropods from the Brazilian Cerrado - a scorpion (*Tityus fasciolatus*), a social wasp (*Polybia sp.*), and a spider (*Acanthoscurria paulensis*). The methods and results are reported and discussed respectively in the Sects. 2 and 3, followed by the conclusion in the Sect. 4.

2   Methods

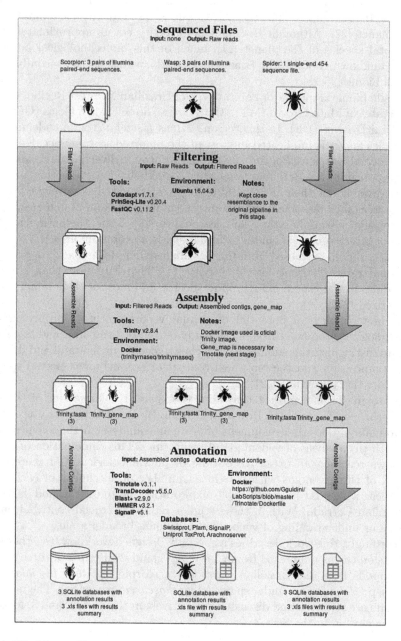

Fig. 1. Workflow - filtering, assembly and annotation - for gland transcripts of representative venom arthropods from the Brazilian Cerrado biome.

Bioinformatics workflows (or pipelines) often have three phases: filtering, assembly, and analysis. The filtering phase consists of quality treatment of the raw data from the sequencer (reads) according to a defined standard. The purpose of the assembly phase is to obtain the largest contigs from the alignment of the reads. Also, the assembly depends on the nature of the data – generally genome or transcriptome – and may or may not use known sequences from the organism itself or another phylogenetically close organism as a reference When the assembly is not reference guided, it is called *de novo* assembly. After the assembly, the performed analyses are diverse and related to the biological query that is proposed to be fulfilled.

In this work, a three-stage bioinformatics workflow was proposed - filtering, assembly, and annotation, which was executed for each organism whose transcripts from the venom glands were sequenced: *Tityus fasciolatus*, *Polybia sp.*, and *Acanthoscurria paulensis*. Figure 1 summarizes the proposed workflow.

2.1 Filtering and Assembly

For scorpion (*Tityus fasciolatus*) and wasp (*Polybia sp.*), transcripts were sequenced in three paired-end runs using Illumina technology, having generated six read sequence files. Spider (*Acanthoscurria paulensis*) had its transcripts sequenced using Roche 454, having produced single-end reads. Reads were filtered using the PRINSEQ [23] software from the reads quality report obtained with the FASTQC [3] software.

After filtering, assembly was performed using the Trinity [12] software. In order to ensure the best possible assembly, Trinity was executed several times for each sequence, using different k-mer sizes between 23 and 31. Table 1 presents a summary of the raw data from the transcriptome sequencing, and the average assembled contigs size, while Fig. 2 presents the assembly stats for Nx^1.

2.2 Annotation

The annotation was firstly performed using Trinotate [13], which is a suite for transcriptome functional annotation that fits non-model organisms. This suite comprises some well-established methods as homology search with Blast+ [2] and SwissProt [8], protein domain identification with HMMER [10] and PFAM [5], and leveraging various annotation databases.

Also, we performed a search for protein signal peptide and transmembrane domain prediction SignalP [4], to use the most recent version of signal peptide database. Moreover, as Trinotate results are stored into an SQLite database, we took advantage of such a feature to perform SQL queries on these data to achieve some relevant biological answers. For example, it was possible to find "GKK" amino acids in the C-terminal portion of peptides annotated as a toxin.

---

[1] It describes the "completeness" of an assembly, where x is a number as 50, which represents the number of contigs covering 50% of the total assembly.

Table 1. Raw data from the transcriptome sequencing, and the average assembled contigs' size.

		Number of reads	Average contigs' size	Number of assembled bases
Tityus fasciolatus	Pair 1	40,000,000	729.09	1,370,087
	Pair 2	40,000,000	421.14	1,359,845
	Pair 3	27,448,123	436.00	1,033,330
Polybia sp.	Pair 1	40,000,000	466.60	1,146,428
	Pair 2	40,000,000	455.77	1,114,368
	Pair 3	27,448,123	449.34	595,825
Acanthoscurria paulensis		607,036	1091.44	15,189,547

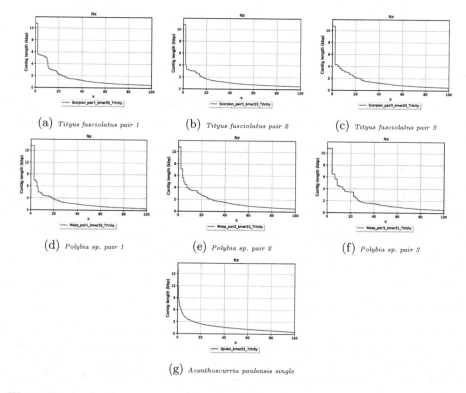

(a) *Tityus fasciolatus pair 1*

(b) *Tityus fasciolatus pair 2*

(c) *Tityus fasciolatus pair 3*

(d) *Polybia sp. pair 1*

(e) *Polybia sp. pair 2*

(f) *Polybia sp. pair 3*

(g) *Acanthoscurria paulensis single*

Fig. 2. Nx plot for the contigs assemblies of *Tityus fasciolatus* pairs 1, 2 and 3; *Polybia sp.* pairs 1, 2, and 3; and *Acanthoscurria paulensis* single-end data.

3   Results and Discussion

Here we describe the results obtained from the application of the designed workflow, which was suitable to the problem of finding venom peptides, and how they were delivered to the stakeholders.

For *Acanthoscurria paulensis*, there were 13,917 assembled transcripts, from which 7,725 were annotated. Regarding metabolism, 3,395 transcripts had annotations related to biological processes, 4,671 related to molecular functions, 2,003 related to cellular components, with 4,340 enzymes. Regarding venom, 347 transcripts were annotated using ToxProt database [15], and 239 using Arachnoserver database [14].

For *Tityus fasciolatus*, there were 8,792 assembled transcripts, from which 4,805 were annotated. Regarding metabolism, 1,107 transcripts had annotations related to biological processes, 1,414 related to molecular functions, 779 related to cellular components, with 1,066 enzymes. Regarding venom, 615 transcripts were annotated using ToxProt database [15], and 195 using Arachnoserver database [14].

For *Tityus fasciolatus*, there were 6,228 assembled transcripts, from which 2,109 were annotated. Regarding metabolism, 880 transcripts had annotations related to biological processes, 1,104 related to molecular functions, 556 related to cellular components, with 2,125 enzymes. Regarding venom, 177 transcripts were annotated using ToxProt database [15], and 68 using Arachnoserver database [14]. A comparative summary of these annotations can be seen in the Fig. 3. Peptides from venoms are complex mixtures of components such as enzymes, metabolites, and many other substances [9]. This arsenal of toxins presents bioactive peptides with potential therapeutic applications [18] and insecticides [26]. Using the Arachnoserver [14] database and the ToxProt database [15], 38 toxins were annotated for *Polybia sp.*, 75 toxins were annotated for *Tityus fasciolatus*, and 117 toxins were annotated for *Acanthoscurria paulensis*.

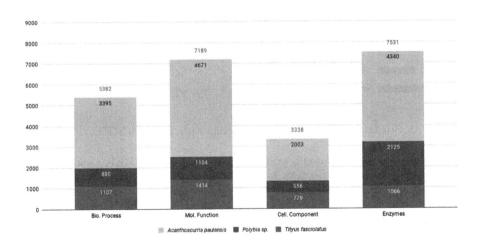

Fig. 3. Comparative amounts for metabolism related annotated transcripts of *Acanthoscurria paulensis*, *Polybia sp.* and *Tityus fasciolatus*.

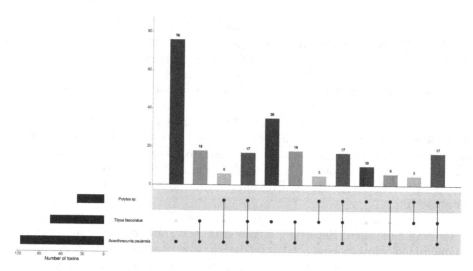

(a) Intersection of annotated peptides.

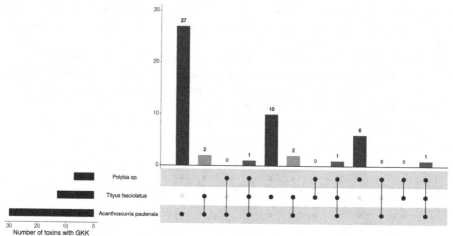

(b) Intersection of annotated peptides presenting "GKK".

Fig. 4. Intersection of annotated peptides, and those presenting "GKK", with ToxProt. Red bars represent unique peptides for each species. Yellow bars represent peptides shared among *Tityus fasciolatus* and *Acanthoscurria paulensis*. Green bars represent peptides shared among *Polybia sp.* and *Acanthoscurria paulensis*. Blue bars represent peptides shared among the three species. (Color figure online)

Venom peptides α-amidation in C-*termini* is a paramount post-translational modification, in most cases essential for their full biological activity recognzing ion-channels, mainly K^+ and Na^+ [6, 9, 18]. As shown in Fig. 4, there are several toxins, unique or shared among the studied arthropods, and the same is valid for those that

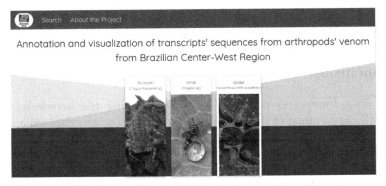

(a) Home page of the Web application.

(b) Annotation of the tree arthropods can be searched.

Fig. 5. Web application to explore data of transcript annotation of three arthropods from the Brazilian Cerrado biome - scorpion (*Tityus fasciolatus*), social wasp (*Polybia sp.*), and spider (*Acanthoscurria paulensis*).

presented "GKK" cleavage site in toxin peptides, which is a well-known marker for the α-amidation [18]. About 35% of the annotated peptides presented the "GKK" cleavage site in *Acanthoscurria paulensis*, 28% in *Tityus fasciolatus*, and 60% in *Polybia sp.*

To organize and provide a unified and helpful repository for the information produced by the pipeline, we have built a Web-based application. This application is consistent with Web standards, allowing the user to access it independently of the operational system and a screen size responsive. Researchers of the Inovatoxin project can perform searches and retrieve data according to their needs, including comparative results.

Similar workflows have been devised for similar tasks, as in Cid-Uribe *et al.* [7]. However, this work brings a semi-automatic built application front-end (Figs. 5a and b), which was developed by leveraging the database generated from the workflow, in which all data is stored. It allows fast development of new applications for similar experiments that use the same pipeline.

4 Conclusion

Aligned with Inovatoxin project's goal of expanding the knowledge of venomous arthropod faune of the Brazilian Cerrado biome, we proposed a specially designed workflow, which resulted in a significant amount of relevant biological data. This data collection is available in an information system to be explored and interpreted, exclusively by the Inovatoxin project researchers.

Data produced from this work allows new insights over representative species' venom components, which have a considerable pharmaceutic and biotechnological potential. The built information systems is a valuable tool to search for unveiled peptides from these Brazilian Cerrado arthropods. Moreover, the successful results obtained from the proposed workflow contributes as a bioinformatics solution for similar studies in other venomous Cerrado's species, providing an even broader understanding of the rich Brazilian Cerrado biome.

The next step is to improve both the database and the information system generated in this step, finding and characterizing enzymes such that we can investigate more deeply the metabolism of these arthropods. After, we intend to study other arthropods from the Brazilian Cerrado biome.

References

1. Abdel-Rahman, M.A., Quintero-Hernández, V., Possani, L.D.: Scorpion venom gland transcriptomics and proteomics: an overview. In: Gopalakrishnakone, P., Calvete, J.J. (eds.) Venom Genomics and Proteomics. T, pp. 105–124. Springer, Dordrecht (2016). https://doi.org/10.1007/978-94-007-6416-3_29
2. Altschul, S.F., Gish, W., Miller, W., Myers, E.W., Lipman, D.J.: Basic local alignment search tool. J. Mol. Biol. **215**(3), 403–410 (1990)
3. Andrews, S., et al.: FastQC: a quality control tool for high throughput sequence data (2010). https://www.bioinformatics.babraham.ac.uk/projects/fastqc/. Accessed July 2019
4. Armenteros, J.J.A., et al.: Signalp 5.0 improves signal peptide predictions using deep neural networks. Nature Biotechnol. **37**(4), 420 (2019)
5. Bateman, A., et al.: The Pfam protein families database. Nucleic Acids Res. **32**(suppl_1), D138–D141 (2004)
6. Chufán, E.E., De, M., Eipper, B.A., Mains, R.E., Amzel, L.M.: Amidation of bioactive peptides: the structure of the lyase domain of the amidating enzyme. Structure **17**(7), 965–973 (2009)
7. Cid-Uribe, J.I., et al.: The diversity of venom components of the scorpion species paravaejovis schwenkmeyeri (scorpiones: Vaejovidae) revealed by transcriptome and proteome analyses. Toxicon **151**, 47–62 (2018)
8. Consortium, T.U.: UniProt: the universal protein knowledgebase. Nucleic Acids Res. **45**(D1), D158–D169 (2016)
9. Delgado-Prudencio, G., Possani, L.D., Becerril, B., Ortiz, E.: The dual α-amidation system in scorpion venom glands. Toxins **11**(7), 425 (2019)

10. Finn, R.D., Clements, J., Eddy, S.R.: Hmmer web server: interactive sequence similarity searching. Nucleic Acids Res. **39**(suppl_2), W29–W37 (2011)
11. de França, E., Vasconcellos, A.G.: Patentes de fitoterápicos no brasil: uma análise do andamento dos pedidos no período de 1995–2017. Cadernos Ciência & Tecnol. **35**(3), 329–359 (2019)
12. Grabherr, M.G., et al.: Full-length transcriptome assembly from RNA-seq data without a reference genome. Nat. Biotechnol. **29**(7), 644 (2011)
13. Haas, B., et al.: Trinotate (2019). https://github.com/Trinotate/Trinotate. Accessed June 2019
14. Herzig, V., et al.: Arachnoserver 2.0, an updated online resource for spider toxin sequences and structures. Nucleic Acids Res. **39**(suppl_1), D653–D657 (2010)
15. Jungo, F., Bougueleret, L., Xenarios, I., Poux, S.: The UniProtKB/swiss-prot toxprot program: a central hub of integrated venom protein data. Toxicon **60**(4), 551–557 (2012)
16. Mortari, M., et al.: Inhibition of acute nociceptive responses in rats after icv injection of thr6-bradykinin, isolated from the venom of the social wasp, Polybia occidentalis. Br. J. Pharmacol. **151**(6), 860–869 (2007)
17. Mourão, C.B.F., et al.: Venomic and pharmacological activity of acanthoscurria paulensis (theraphosidae) spider venom. Toxicon **61**, 129–138 (2013)
18. Ortiz, E., Gurrola, G.B., Schwartz, E.F., Possani, L.D.: Scorpion venom components as potential candidates for drug development. Toxicon **93**, 125–135 (2015)
19. Prezoto, F., de Souza, M.M., Elpino-Campos, A., Del-Claro, K., et al.: New records of social wasps (hymenoptera, vespidae) in the Brazilian tropical savanna. Sociobiology **54**(3), 759 (2009)
20. Pucca, M.B., Oliveira, F.N., Schwartz, E.F., Arantes, E.C., Lira-da Silva, R.M.: Scorpionism and dangerous species of Brazil. In: Gopalakrishnakone, P. (ed.) Toxinology, pp. 1–24. Springer, Dordrecht (2021). https://doi.org/10.1007/978-94-007-6647-1_20-1
21. Reumont, V., Marcus, B., Drukewitz, S.H.: The significance of comparative genomics in modern evolutionary venomics. Fronti. Ecol. Evol. **7**, 163 (2019)
22. Romero-Gutiérrez, M., Santibáñez-López, C., Jiménez-Vargas, J., Batista, C., Ortiz, E., Possani, L.: Transcriptomic and proteomic analyses reveal the diversity of venom components from the vaejovid scorpion serradigitus gertschi. Toxins **10**(9), 359 (2018)
23. Schmieder, R., Edwards, R.: Quality control and preprocessing of metagenomic datasets. Bioinformatics **27**(6), 863–864 (2011)
24. Schwartz, E.N.F.: Rede pró-centro-oeste (2019). http://200.129.206.69/projects. Accessed July 2019
25. Silva, S.D.S., Azevedo, G.G., Silveira, O.T.: Social wasps of two cerrado localities in the northeast of maranhão state, Brazil (hymenoptera, vespidae, polistinae). Rev. Bras. Entomol. **55**(4), 597–602 (2011)
26. Smith, J.J., Herzig, V., King, G.F., Alewood, P.F.: The insecticidal potential of venom peptides. Cell. Mol. Life Sci. **70**(19), 3665–3693 (2013). https://doi.org/10.1007/s00018-013-1315-3
27. Tedford, H.W., Sollod, B.L., Maggio, F., King, G.F.: Australian funnel-web spiders: master insecticide chemists. Toxicon **43**(5), 601–618 (2004)
28. Zappi, D.C., et al.: Growing knowledge: an overview of seed plant diversity in Brazil. Rodriguésia **66**(4), 1085–1113 (2015)

Block-Interchange Distance Considering Intergenic Regions

Ulisses Dias[1(✉)] ⓘ, Andre Rodrigues Oliveira[2] ⓘ, Klairton Lima Brito[2] ⓘ,
and Zanoni Dias[2] ⓘ

[1] School of Technology, University of Campinas, Limeira, Brazil
ulisses@ft.unicamp.br
[2] Institute of Computing, University of Campinas, Campinas, Brazil
{andrero,klairton,zanoni}@ic.unicamp.br

Abstract. Genome Rearrangement (GR) is a field of computational
biology that uses conserved regions within two genomes as a source
of information for comparison purposes. This branch of genomics uses
the order in which these regions appear to infer evolutive scenarios and
to compute distances between species, while usually neglecting non-
conserved DNA sequence. This paper sheds light on this matter and
proposes models that use both conserved and non-conserved sequences
as a source of information. The questions that arise are how classic GR
algorithms should be adapted and how much would we pay in terms
of complexity to have this feature. Advances on these questions aid in
measuring advantages of including such approach in GR algorithms. We
propose to represent non-conserved regions by their lengths and apply
this idea in a genome rearrangement problem called "Sorting by Block-
Interchanges". The problem is an interesting choice on the theory of
computation viewpoint because it is one of the few problems that are
solvable in polynomial time and whose algorithm has a small number of
steps. That said, we present a 2-approximation algorithm to this problem
along with data structures and formal definitions that may be generalized
to other problems in GR field considering intergenic regions.

Keywords: Comparative genomics · Genome rearrangements ·
Intergenic regions

1 Introduction

Genomes undergo large-scale events called rearrangements during the evolutionary process. To infer which rearrangements have occurred in the past, one needs
a global view of genomes which is possible when whole-genome data exists for
a given species. Such inference uses the principle of parsimony to compute the
most likely scenario, formulated as the minimum number of large-scale events
that explain observed differences.

In the Genome Rearrangement (GR) field, genomes are represented by the
order of conserved blocks in it. Given two closely related species, a high coverage

© Springer Nature Switzerland AG 2020
L. Kowada and D. de Oliveira (Eds.): BSB 2019, LNBI 11347, pp. 58–69, 2020.
https://doi.org/10.1007/978-3-030-46417-2_6

of conserved regions is expected. Regions for which no counterpart is found in other genomes are overlooked in the process.

In the GR field, Sorting by Transpositions [1] and Sorting by Reversals [10] are well studied problems. The former considers events that swap the position of two adjacent segments inside a genome, and the latter considers events that reverse the order and the orientation of a DNA sequence. Double-Cut and Join (DCJ) [12] is also a GR problem. DCJ events cut a genome between adjacent blocks a and b, and adjacent blocks c and d, and joins either a to c and b to d, or a to d and b to c.

In most cases, conserved regions are genes whose destruction would impact on organism's survival. Regions between those genes are called intergenic regions, and they are usually less conserved because the impact on survival is minor. We hereafter use the term gene to indicate conserved regions, and intergenic region to indicated non-conserved DNA sequences. Studies indicate that incorporating the sizes of intergenic regions in GR models make the distance estimations match more closely the evolutionary history in experiments *in silico* [2,3].

Genomes also undergo non-conservative events that alter the amount of genetic material such as insertions and deletions. The first adds genetic material to the genome and the latter removes genetic material from the genome.

Results considering intergenic sizes for models with DCJs and DCJs together with insertions and deletions are known: the former is NP-hard and it has a 4/3-approximation algorithm [8], while the latter is polynomial [5]. Besides, approximation algorithms considering intergenic sizes with super short reversals (a reversal acting on at most two genes) on unsigned permutations were recently published [11].

To bridge the gap between classic GR problems – like the Sorting by Reversals and Sorting by Transpositions problems – and comparison models that use several features of the genome as source of information, we represent intergenic regions by their sizes and add this information to the so-called Cycle Graph structure, a tool widely used to solve GR problems. That allows to represent both genes and intergenic regions in the same graph structure, a first step to derive algorithms.

The next question is to choose a GR problem to work on. The Sorting by Transpositions problem is NP-hard [4] and the best algorithm has an approximation factor of 1.375 [7]. The Sorting by Reversals problem has a polynomial time solution when the orientation of the genes is considered [9], but the original algorithm that uses the Cycle Graph structure is far from trivial and it is full of intermediate steps. Adding intergenic region sizes in a complicated problem could lead to a cumbersome algorithm, not the purpose of this paper.

A Block-Interchange (BI) is an event that swaps positions of two blocks located anywhere in the genome. The Sorting by BIs problem has a relatively simple polynomial algorithm that uses the Cycle Graph as a data structure [6]. We added the concept of intergenic region size to this problem and developed a 2-approximation algorithm. We also added the non-conservative events of insertions and deletions of intergenic material and kept the same approximation ratio.

Let us first briefly introduce the problem and how BIs behave. Figure 1 depicts the events we want to represent. Blocks labeled as `Gene 1`, `Gene 2`, and so on, are conserved regions. We place letters `A`, `C`, `G`, and `T` to represent a sequence of intergenic regions. Letters are provided for illustrative purposes only, as placeholders, since we will later represent intergenic regions by their sizes.

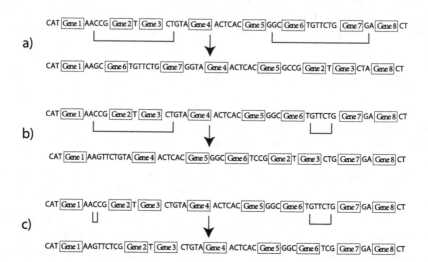

Fig. 1. Block-interchanges (BIs): (**a**) shows two blocks carrying genes, (**b**) shows only one block with genes, and (**c**) shows two blocks without genes.

Figure 1(a) presents a BI swapping two blocks with genes surrounded by intergenic regions. Genes are indivisible, so we can identify their positions after a BI takes place, whereas intergenic regions change after a rearrangement event and could no longer be identified as the same as before. That explains why we represent intergenic regions solely by their lengths. Figure 1(b) presents a BI such that one of the blocks has no gene, which means it moves only intergenic regions around. Figure 1(c) presents a BI having blocks without genes. In this case, the operation makes sense only if the sizes of the blocks are different, so the order of the genes remains, but the distance between genes in terms of intergenic region sizes does not.

This work is organized as follows. Section 2 formally defines GR problems that consider intergenic regions, and introduces general definitions suitable for any problem of the kind. Section 3 describes the weighted graph structure useful for many GR problems. Section 4 formally defines the BI problem that considers intergenic regions. Section 5 makes use of knowledge in previous sections to derive a 2-approximation algorithm. Section 6 generalizes our model to include insertions and deletions. Section 7 concludes the paper.

2   Formal Definition of Intergenic Regions

A genome $\mathcal{G}$ is a sequence of genes $(\pi_1, \pi_2, \ldots, \pi_n)$ alternated with a sequence of intergenic regions $(\breve{\pi}_1, \breve{\pi}_2, \ldots, \breve{\pi}_{n+1})$: $\mathcal{G} = \breve{\pi}_1, \pi_1, \breve{\pi}_2, \pi_2, \ldots, \breve{\pi}_n, \pi_n, \breve{\pi}_{n+1}$. We assume two genomes share the same set of genes, which may appear in different orders due to genome rearrangements, and there exists no duplicated genes. That said, we choose one of the genomes as a reference and label their genes with natural numbers according to the order of appearance. In Fig. 2, the set of genes in $\mathcal{G}_1$ would be labeled as $1, 2, \ldots, 8$. Genes in the second genome keep the label, hence, $\mathcal{G}_2$ gene order is represented by the permutation $\pi = (3\ 2\ 1\ 7\ 6\ 4\ 8\ 5)$.

The goal of most GR problems is to transform one genome into the other with the minimum number of operations. Genome $\mathcal{G}_1$ is usually labeled as previously described, which leads to the so-called identity permutation $\iota = (1\ 2\ \ldots\ n)$ for a genome having n genes. Therefore, problems in the field aim at sorting permutations $\pi = (\pi_1\ \pi_2\ \ldots\ \pi_n)$, $\pi_i \in \mathbb{N}$, $1 \leq \pi_i \leq n$, and $\pi_i \neq \pi_j$ for all $i \neq j$.

We propose a more elaborated class of problems by adding intergenic regions. We represent intergenic regions by their lengths instead of assigning unique identifiers to each of them. The sequence of intergenic regions around n genes is represented as $\breve{\pi} = (\breve{\pi}_1\ \breve{\pi}_2\ \ldots\ \breve{\pi}_{n+1})$, $\breve{\pi}_i \in \mathbb{N}$. Intergenic region $\breve{\pi}_i$ is on the left side of π_i, whereas $\breve{\pi}_{i+1}$ is on the right side.

Figure 2 aids once again to understand the concepts. We assume the number of nucleotides between genes is a good estimator for intergenic region length. For example, in $\mathcal{G}_1$ before "Gene 1" we have 3 nucleotides, between "Gene 1" and "Gene 2" we have 5, and so on, as shown in Fig. 2(b). Thus, $\pi = (3\ 2\ 1\ 7\ 6\ 4\ 8\ 5)$ represents gene orders, while $\breve{\pi} = (1\ 7\ 0\ 4\ 2\ 7\ 4\ 0\ 9)$, and $\breve{\iota} = (3\ 5\ 1\ 5\ 6\ 3\ 7\ 2\ 2)$ represent intergenic regions from both genomes. No other information is retrieved, so an instance of our problem aggregates $(\pi, \breve{\pi}, \breve{\iota})$, such that $\sum_{i=1}^{n+1} \breve{\pi}_i = \sum_{i=1}^{n+1} \breve{\iota}_i$, which guarantees that the total intergenic region lengths are conserved. This last constraint may be relaxed as long as one allows non-conservative mutations. We analyze this case in Sect. 6.

To formally define the class of GR problems that considers intergenic regions, we need to establish the goal. Let us call a *model* $\mathcal{M}$ a set of genome rearrangements that can be used to change a permutation and the intergenic regions. Therefore, for $\rho \in \mathcal{M}$, the notation $(\pi, \breve{\pi}) \cdot \rho = (\sigma, \breve{\sigma})$ indicates the outcome of ρ, which transforms π into σ, and $\breve{\pi}$ int $\breve{\sigma}$. The goal is to compute the shortest sequence of events from $\mathcal{M}$ that transforms π into ι, and $\breve{\pi}$ into $\breve{\iota}$. Therefore, the distance $d_{\mathcal{M}}(\pi, \breve{\pi}, \breve{\iota}) = m$ implies a minimal sequence of operations $\rho_1, \rho_2, \ldots, \rho_m$ such that $(\pi, \breve{\pi}) \cdot \rho_1 \cdot \ldots \cdot \rho_m = (\iota, \breve{\iota})$.

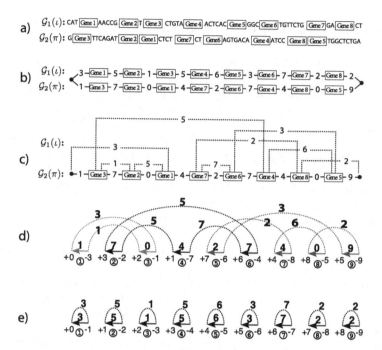

Fig. 2. We illustrate how two genomes in (**a**) could be used to represent intergenic regions by their sizes in (**b**), and later joined together in the same diagram in (**c**). The Weighted Cycle Graph structure in (**d**) carries all the information we need from the original genomes and may be properly handled to reach the goal shown in (**e**).

3   The Weighted Cycle Graph Structure

We adapted a graph structure called *Cycle Graph* [1,9] to represent an instance $(\pi, \breve{\pi}, \breve{\iota})$ in a single graph. Before jumping to the definition, let us recall the example in Fig. 2 and see how the graph representation naturally follows. In (a) we see two genomes as a sequence of genes and intergenic regions. In (b) we replace intergenic regions by their lengths. In (c) we show both genomes in the same diagram, such that each gene appears only once. We represent π and $\breve{\pi}$ as a horizontal lines and $\breve{\iota}$ as dashed lines above. In (d) we show an *Weighted Cycle Graph*. Observe that it comes directly from the diagram in (c), we insert two vertices $(-i, +i)$ for each "Gene i", $1 \le i \le 8$, and we also add $+0$ and -9 in each extremity as landmarks for the start and end points.

Formally, the *Weighted Cycle Graph* $G(\pi, \breve{\pi}, \breve{\iota}) = (V, E, w)$, where V is the set $\{-n, \ldots, -2, -1, +1, +2, \ldots, +n\} \cup \{+0, -(n+1)\}$, E is the set of edges, and $w : E \to \mathbb{N}$ is a function mapping edges to values corresponding to intergenic region length. Edges can be called *black*: $\{e_i = (-\pi_i, +\pi_{i-1}) : 1 \le i \le n+1\}$, and $w(e_i) = \breve{\pi}_i$; or *gray*: $\{e'_i = (+(i-1), -i) : 1 \le i \le n+1\}$, and $w(e'_i) = \breve{\iota}_i$. Black edges are the solid horizontal bottom lines in Fig. 2(d), while gray edges are the dashed arcs. In this definition, we also consider $\pi_0 = 0$ and $\pi_{n+1} = n+1$.

Each vertex in $G(\pi, \breve{\pi}, \breve{\iota})$ has a gray edge and a black edge incident to it, which allows a unique decomposition of edges in cycles of alternating colors. In Fig. 2(d), we have three cycles. We label black edge e_i as i, and in Fig. 2(d) we place labels as circumscribed numbers below black edges. Labels allow us to identify a cycle as the list of black edges reached when we traverse it. To make the notation unique, we start traversing cycles by the black edge with the largest label. For example, in Fig. 2(d) we have three cycles: $(9, 5, 7, 8)$, $(6, 4, 2)$, and $(3, 1)$.

A cycle $C = (c^1, \ldots, c^\ell)$ is *called non-oriented* if $c^1, \ldots, c^\ell$ is a decreasing sequence, as in $(6, 4, 2)$ and $(3, 1)$; C is *oriented* otherwise, as in $(9, 5, 7, 8)$. A cycle $C = (c^1, \ldots, c^\ell)$ is called *balanced* if $\sum_{i=1}^{l}[w(e'_{c^i}) - w(e_{c^i})] = 0$, and is *unbalanced* otherwise. That is, one cycle is balanced if the sum of weights of gray edges equals the sum of weights of black edges, and it is unbalanced otherwise.

An unbalanced cycle is *positive* if the sum of weights in gray edges is greater than the black edge counterpart, and it is *negative* otherwise. In Fig. 2(d), the cycle $(3, 1)$ is positive since gray edges sum up to $3 + 1 = 4$, whereas black edges sum up to $0 + 1 = 1$. Cycle $(6, 4, 2)$ is negative since gray edges sum up to $7 + 5 + 5 = 17$ whereas black edges sum up to $7 + 7 + 4 = 18$.

We use the notations $c(\pi, \breve{\pi}, \breve{\iota})$ and $c_b(\pi, \breve{\pi}, \breve{\iota})$ to denote the number of cycles and balanced cycles in $G(\pi, \breve{\pi}, \breve{\iota})$, respectively. The goal of GR algorithms is to reach the identity permutation and to change the lengths in $\breve{\pi}$ so that it equals to $\breve{\iota}$. In other words, the goal is to reach the genome $(\iota, \breve{\iota})$. An example of such configuration is depicted in Fig. 2(e). Observe that all cycles have only one black edge and only one gray edge, so the number of cycles is $n + 1$, such that n is the number of elements in π. In addition, each cycle is balanced.

As a general rule, given an input instance $(\pi, \breve{\pi}, \breve{\iota})$ and a model $\mathcal{M}$, algorithms should increase the number of balanced cycles from $c_b(\pi, \breve{\pi}, \breve{\iota})$ to $n + 1$ using operations in $\mathcal{M}$. For a BI ρ such that $(\pi, \breve{\pi}) \cdot \rho = (\sigma, \breve{\sigma})$, let $\Delta c_b(\pi, \breve{\pi}, \breve{\iota}, \rho) = c_b(\sigma, \breve{\sigma}, \breve{\iota}) - c_b(\pi, \breve{\pi}, \breve{\iota})$ denote the variation in the number of balanced cycles when $\rho \in \mathcal{M}$ is applied, and let $\Delta_{\mathcal{M}}$ be the maximum value of $\Delta c_b(\pi, \breve{\pi}, \breve{\iota}, \rho)$, for all possible $(\pi, \breve{\pi}, \breve{\iota})$, and for all $\rho \in \mathcal{M}$. Theorem 1 is a lower bound feasible for any GR problem that can be represented using the Weighted Cycle Graph structure.

Theorem 1. *For any instance* $(\pi, \breve{\pi}, \breve{\iota})$, $d_{\mathcal{M}}(\pi, \breve{\pi}, \breve{\iota}) \geq \frac{n + 1 - c_b(\pi, \breve{\pi}, \breve{\iota})}{\Delta_{\mathcal{M}}}$.

Proof. We know that $c_b(\iota, \breve{\iota}, \breve{\iota}) = n + 1$. Therefore, the goal is to increase the number of cycles from $c_b(\pi, \breve{\pi}, \breve{\iota})$ to $n+1$; since this number increases by at most $\Delta_{\mathcal{M}}$ for each operation ρ in $\mathcal{M}$, the theorem follows.   □

4   The Block-Interchange Problem

Let us clarify how a BI affects Weighted Cycle Graph structures. Consider the effect of applying the 3 BIs illustrated in Fig. 1 on the Weighted Cycle Graph, as shown in Fig. 3. Some properties should be noted:

1. BIs may act on two, three or four black edges.
2. Affected black edges may be split once or twice, and weights are distributed.

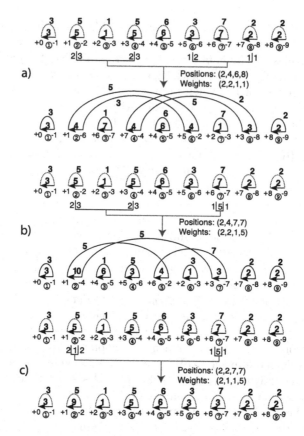

Fig. 3. Examples of block-interchanges impacting Weighted Cycle Graph structures. In (a) we show an outcome of applying a BI in four black edges. In (b) we show a BI applied in three black edges, and in (c) a BI is a applied in two black edges.

3. We need eight values to characterize a BI: four indices for the positions of black edges it acts on, and four values to redistribute weights. Black edges represent intergenic regions that may be split in two or three segments. We identify the length of the left segment if a black edge is split once; and we use lengths of the left and the inner segments if a black edge is split twice.

Christie [6] sets a limit to the increase in the number of cycles after a BI takes place: no block-interchange increases $c(\pi, \breve{\pi}, \breve{\iota})$ by more than 2. We use it to limit the change in the number of balanced cycles.

Lemma 1. *No block-interchange increases $c_b(\pi, \breve{\pi}, \breve{\iota})$ by more than 2.*

Proof. Christie [6] tells we can increase the number of cycles by at most 2. Two scenarios are possible: either (i) one cycle is split in three, or (ii) two cycles are split in four by a single block-interchange ρ.

Let us consider the first scenario, and let C be the cycle split in three. If C is balanced, the best we expect is that ρ creates three balanced cycles, so $\Delta c_b(\pi, \breve{\pi}, \breve{\iota}, \rho) \leq 2$. Otherwise, at least one of the resulting cycles shall be unbalanced, since weights of black edges of the three cycles sum up to a value different from the sum of gray edge weights. Therefore, the best we expect is to create at most 2 balanced cycles, so again $\Delta c_b(\pi, \breve{\pi}, \breve{\iota}, \rho) \leq 2$.

Let us consider the second scenario, and let C and D be the cycles split by the BI. C generates C' and C'', while D generates D' and D''. If C is balanced, C' and C'' could be both balanced, but if C is unbalanced, only one of them can be balanced. Therefore the change is by at most 1. The same applies for D and the cycles generated by the BI (D' and D''). Therefore, a BI can use both C and D to generate at most 2 new balanced cycles, so $\Delta c_b(\pi, \breve{\pi}, \breve{\iota}, \rho) \leq 2$.   □

Let us call $\mathcal{BI}$ the model that allows only block-interchanges, Theorem 2 states a lower bound.

Theorem 2. $d_{\mathcal{BI}}(\pi, \breve{\pi}, \breve{\iota}) \geq \frac{n+1-c_b(\pi,\breve{\pi},\breve{\iota})}{2}$.

Proof. Straightforward from Theorem 1 and Lemma 1.

5   A 2-Approximation Algorithm

We present an algorithm that guarantees an approximation ratio of 2. The algorithm first searches for negative cycles, and we show that if they exist it is always possible to increase the number of balanced cycles by 1 using BIs (Lemmas 2, 3, 4). If no negative cycle can be found, then no positive cycle can be found either. In that case, we break balanced cycles with more than one black edge into smaller balanced ones.

A cycle is trivial if it has 1 black edge; otherwise it is non-trivial. A non-trivial cycle $C = (c^1, \ldots, c^\ell)$ can be either oriented or non-oriented. An oriented cycle has at least one triple of black edges c^i, c^j, and c^k such that $i > j > k$ and $c^i > c^k > c^j$ [1]. A triple of such form is called an oriented triple, and it cannot be found in non-oriented cycles.

Lemma 2. *Let C be a trivial negative cycle, it is possible to increase the number of balanced cycles by 1 using a block-interchange.*

Proof. Since $C = (c^1)$ is unbalanced, there exists another unbalanced cycle D in the graph. Therefore, we can apply a block-interchange to move the extra weight from C to D and create a balanced cycle. Figure 4(a) illustrates this lemma.   □

Lemma 3. *Let C be a non-trivial non-oriented cycle, if C is either balanced or negative, it is possible to increase the number of balanced cycles by 1 using a block-interchange.*

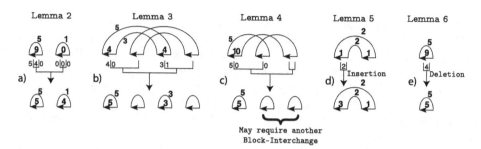

Fig. 4. Example for each step of the algorithms.

Proof. If the cycle $C = (c^1, \ldots, c^\ell)$ is non-oriented, Bafna and Pevzner [1] showed that for every e_{c^i} and e_{c^j} from C with $c^i > c^j$ there exists another cycle D with black edges e_{d^i} and e_{d^j} such that either $c^i > d^i > c^j > d^j$ or $d^i > c^i > d^j > c^j$. These edges are where C and D intersect. A block-interchange applied on these black edges will create four cycles C', C'', D', and D'' such that C' is formed by the path that goes from e_{c^i} to e_{c^j} – excluding e_{c^i} and e_{c^j} – with a new black edge that can receive its weight from both e_{c^i} and e_{c^j}.

To make C' balanced, we need to find black edges e_{c^i} and e_{c^j} such that the path that goes from one to the other has the sum of weights of black edges greater than or equal to the sum of weights of gray edges. That will give us room to adjust the extra weight and send it to C''. Note that the number of black edges in this path is greater than the number of gray edges by 1. Since C is either balanced or negative, it is always possible to find e_{c^i} and e_{c^j}.

In case of C negative, we create C' balanced which increases the number of balanced cycles by 1. In case of C balanced, we create C' balanced which will force C'' to be balanced too, so the number of balanced cycles also increases by 1. Figure 4(b) shows a balanced cycle being broken into two balanced ones.   □

Lemma 4. *Let C be an oriented cycle, if C is either balanced or negative, it is possible to increase the number of balanced cycles by 2 using at most two block-interchanges.*

Proof. Let $C = (c^1, \ldots, c^\ell)$ be the oriented cycle and (c^x, c^y, c^z) be the oriented triple. We can apply a block-interchange on these black edges c^x, c^y, and c^z that creates three cycles C', C'', and C'''. Observe that one of them will be split twice by the block-interchange.

Without loss of generality, we assume that C' is formed by the path that goes from e_{c^z} to e_{c^y} with a new black edge that can receive its weight from both e_{c^y} and e_{c^z}, C'' is formed by the path that goes from e_{c^x} to e_{c^z} with a new black edge that can receive its weight from both e_{c^x} and e_{c^z}, and C''' is formed by the path that goes from e_{c^y} to e_{c^x} with a new black edge that can receive its weight from both e_{c^y} and e_{c^x}. Since C is either balanced or negative, we can further constrain one of the black edges in the triple to have its weight greater than or equal to the weight of its left gray edge. That said, we apply the BI

Algorithm 1: A 2-approximation algorithm for block-interchanges.

---

1 $distance \leftarrow 0$
2 while *the current Weighted Cycle Graph G is different from* $G(\iota, \check{\iota}, \check{\iota})$ **do**
3   **if** *a trivial negative cycle C exists in G* **then**
4   Apply a block-interchange from Lemma 2 and update G.
5 $distance \leftarrow distance + 1$
6   **else if** *a non-oriented balanced or negative cycle C exists in G* **then**
7   Apply a block-interchange from Lemma 3 and update G.
8 $distance \leftarrow distance + 1$
9   **else if** *a oriented balanced or negative cycle C exists in G* **then**
10   Apply k block-interchanges from Lemma 4 and update G.
11 $distance \leftarrow distance + k$

12 return $distance$

---

such that one of the generated cycles is balanced, which requires to distribute the extra-weight from e_{c^x}, e_{c^y}, and e_{c^z}, as illustrated by Fig. 4(c).

Assume that C' is the balanced cycle after the first block-interchange. If one of the other cycles is balanced, nothing needs to be done and the lemma follows. However, if both C'' and C''' are unbalanced, one of them is negative (since C is either balanced or negative). Therefore, we either use Lemma 2 if the negative cycle is trivial, or Lemma 3 if it is non-oriented, or we find another oriented triple to apply a second block-interchange if it is oriented.   □

Algorithm 1 groups the lemmas together. Each step guarantees either a new balanced cycle after one block-interchange (Lemmas 2 and 3) or two balanced cycles after two block-interchanges (Lemma 4). By Lemma 1 at most two balanced cycles can be created after one operation, so the approximation factor of 2 follows. Algorithm 1 runs in $O(n^3)$ since at each step we iterate over the Weighted Cycle Graph once to find the proper cycle, we iterate over the cycle once to select the block-interchange, and the distance is lower than or equal to $n + 1$.

6   Considering Insertions and Deletions

In previous sections, our model constrains genomes to have the same overall size of intergenic region, which may be unfeasible. We now relax this constraint by allowing two non-conservative operations: insertions and deletions. Both operations apply to intergenic material only, i.e., insertions increase the size of an intergenic region, while deletions decrease the size of an intergenic region.

Lemma 2 is the only lemma from Algorithm 1 that is no longer valid when using insertions and deletions. That happens because it requires at least two unbalanced cycles, which we can no longer guarantee. Lemmas 3 and 4 are still valid and deal with cases where we have either negative or balanced non-trivial

---

Algorithm 2: A 2-approximation algorithm for block-interchanges, insertions, and deletions.

```
1  distance ← 0
2  while the current Weighted Cycle Graph G is different from G(ι, ῐ, ῐ) do
3  │   if a non-oriented balanced or negative cycle C exists in G then
4  │   │   Apply a block-interchange from Lemma 3 and update G.
5  │   └   distance ← distance + 1
6  │   else if a oriented balanced or negative cycle C exists in G then
7  │   │   Apply k block-interchanges from Lemma 4 and update G.
8  │   └   distance ← distance + k
9  │   else if a positive cycle C exists in G then
10 │   │   Apply an insertion from Lemma 5 and update G.
11 │   └   distance ← distance + 1
12 │   else if a trivial negative cycle C exists in G then
13 │   │   Apply a deletion from Lemma 6 and update G.
14 └   └   distance ← distance + 1
15 return distance
```

---

cycles. We use insertions and deletions to cases that are not covered, as shown in Algorithm 2.

Assuming that the graph has only positive cycles, it is possible to create one balanced cycle using one insertion (Lemma 5). In cases the graph has (one or more) trivial negative cycles, it is possible to create one balanced cycle using one deletion (Lemma 6).

Lemma 5. *Let C be a positive cycle, it is possible to increase the number of balanced cycles by 1 using one insertion.*

Proof. We add in the first black edge of C the weight required to turn it balanced. Figure 4(d) illustrates this lemma.   □

Lemma 6. *Let C be a trivial negative cycle, it is possible to increase the number of balanced cycles by 1 using one deletion.*

Proof. We remove from the only black edge in C the extra weight. Figure 4(e) illustrates this lemma.   □

Algorithm 2 adds the new lemmas. Note that both Lemma 5 and Lemma 6 increase the number of balanced cycles by 1 after an operation, which means the approximation factor 2 remains. The algorithm still runs in $O(n^3)$ by the same arguments as before.

7   Conclusion

This paper proposes a class of genome rearrangement problems that considers the gene order and the sizes of intergenic regions. We adapted classic tools to serve as a starting point to design lower bounds and approximations.

We believe this work offers insights to set the stage for other genome rearrangement problems that may benefit from considering intergenic regions, such as those that use events like reversals and transpositions, or even a combination of several events. That would increase the biological appeal of genome rearrangement problems. Algorithms for these new problems along with their proper formalization and NP-hardness proofs are left for future works.

Acknowledgments. This work was supported by the National Council for Scientific and Technological Development - CNPq (grants 400487/2016-0, 425340/2016-3, 304380/2018-0, and 140466/2018-5), the São Paulo Research Foundation - FAPESP (grants 2015/11937-9, 2017/12646-3, and 2017/16246-0), the Brazilian Federal Agency for the Support and Evaluation of Graduate Education - CAPES, and the CAPES/COFECUB program (grant 831/15).

References

1. Bafna, V., Pevzner, P.A.: Sorting by transpositions. SIAM J. Disc. Math. **11**(2), 224–240 (1998). https://doi.org/10.1137/S089548019528280X
2. Biller, P., Guéguen, L., Knibbe, C., Tannier, E.: Breaking good: accounting for fragility of genomic regions in rearrangement distance estimation. Genome Biol. Evol. **8**(5), 1427–1439 (2016). https://doi.org/10.1093/gbe/evw083
3. Biller, P., Knibbe, C., Beslon, G., Tannier, E.: Comparative genomics on artificial life. In: Beckmann, A., Bienvenu, L., Jonoska, N. (eds.) CiE 2016. LNCS, vol. 9709, pp. 35–44. Springer, Cham (2016). https://doi.org/10.1007/978-3-319-40189-8_4
4. Bulteau, L., Fertin, G., Rusu, I.: Sorting by transpositions is difficult. SIAM J. Comput. **26**(3), 1148–1180 (2012). https://doi.org/10.1137/110851390
5. Bulteau, L., Fertin, G., Tannier, E.: Genome rearrangements with indels in intergenes restrict the scenario space. BMC Bioinform. **17**(S14), 225–231 (2016). https://doi.org/10.1186/s12859-016-1264-6
6. Christie, D.A.: Sorting permutations by block-interchanges. Inf. Process.Lett. **60**(4), 165–169 (1996)
7. Elias, I., Hartman, T.: A 1.375-approximation algorithm for sorting by transpositions. IEEE/ACM Trans. Comput. Biol. Bioinform. **3**(4), 369–379 (2006)
8. Fertin, G., Jean, G., Tannier, E.: Algorithms for computing the double cut and join distance on both gene order and intergenic sizes. Algorithms Mol. Biol. **12**(16), 1–11 (2017). https://doi.org/10.1186/s13015-017-0107-y
9. Hannenhalli, S., Pevzner, P.A.: Transforming men into mice (polynomial algorithm for genomic distance problem). In: Proceedings of 36th Annual IEEE Symposium on Foundations of Computer Science (FOCS 1995), pp. 581–592. IEEE Computer Society Press, Washington, DC (1995). https://doi.org/10.1109/SFCS.1995.492588
10. Keccioglu, J.D., Sankoff, D.: Exact and approximation algorithms for sorting by reversals, with application to genome rearrangement. Algorithmica **13**, 180–210 (1995). https://doi.org/10.1007/BF01188586
11. Oliveira, A.R., Jean, G., Fertin, G., Dias, U., Dias, Z.: Super short reversals on both gene order and intergenic sizes. In: Alves, R. (ed.) BSB 2018. LNCS, vol. 11228, pp. 14–25. Springer, Cham (2018). https://doi.org/10.1007/978-3-030-01722-4_2
12. Yancopoulos, S., Attie, O., Friedberg, R.: Efficient sorting of genomic permutations by translocation, inversion and block interchange. Bioinformatics **21**(16), 3340–3346 (2005). https://doi.org/10.1093/bioinformatics/bti535

K-mer Mapping and RDBMS Indexes

Elvismary Molina de Armas[1][(✉)], Paulo Cavalcanti Gomes Ferreira[3],
Edward Hermann Haeusler[1], Maristela Terto de Holanda[2],
and Sérgio Lifschitz[1]

[1] Depto. Informática, PUC-Rio, Rio de Janeiro, Brazil
{earmas,hermann,sergio}@inf.puc-rio.br
[2] Depto. Ciência da Computação, UNB, Brasília, Brazil
mholanda@unb.br
[3] Depto. Bioquimica Médica, UFRJ, Rio de Janeiro, Brazil
paulof@bioqmed.ufrj.br

Abstract. K-mer Mapping, an internal process for De Novo NGS
genome fragments assembly methods, constitutes a computational challenge due to its high main memory consumption. We present a study of
index-based methods to deal with this problem, considering a RDBMS
environment. We propose an ad-hoc I/O cost model and analyze the
performance of hash and B-tree versions for index structures. Furthermore, we present a novel approach for an index based on hashing that
takes into account the notion of minimum substrings. An actual RDBMS
implementation for experiments with a sugarcane dataset shows that one
can obtain considerable performance gains while reducing main memory
requirements.

1 Introduction

The computational fragment assembly [8] is still a current fundamental problem
for bioinformatics. For *de novo* [18] assembling, with no reference genome to
guide the process. To address this problem, several genome assemblers have been
implemented based on a *de Bruijn* graph structure [2,13,20,21]. Assemblathon
[1,7] and Gage [17] present a preliminary evaluation of these assemblers.

However, constructing the *de Bruijn* graph approach and its processing,
require an enormous amount of memory - much more than most affordable off-
the-shelf computers - which is a significant drawback. Experiments such as [4,11]
and supplementary results in [1] confirm that the *de Bruijn* graph construction
has the highest rate of memory and runtime consumption in those assemblers.

The first step in the construction of a *de Bruijn* graph is decomposing the
DNA fragments, also called reads, into *k-mers*, which become the nodes of the
graph. A *k-mer* is a k length substring of a read. Then an edge is set between
two nodes if the *k-mers* of those nodes occur consecutively in at least one read.
Since there can be repeated *k-mers*, duplicated *k-mers* should be mapped to
the same node, known as K-mer Mapping. Once the maps are ready, to create

© Springer Nature Switzerland AG 2020
L. Kowada and D. de Oliveira (Eds.): BSB 2019, LNBI 11347, pp. 70–82, 2020.
https://doi.org/10.1007/978-3-030-46417-2_7

the (directed) edge set for the *de Bruijn* graph the short reads are scanned. Therefore, the K-mer Mapping process defines the main memory requirements.

Relational Database Management Systems (RDBMS) have been designed and optimized to work with a massive volume of data and limited main memory. They are adjusted to minimize disk I/O operations using strategies like indexes and buffer manager algorithms. The use of RDBMS for implementing the K-mer Mapping was proposed in [5]. An RDBMS has an abstraction layer that optimizes the execution of disk persistence and disk-access, reducing memory consumption by using proven persistent data structures and systems.

This paper studies the impact of different index-based procedures to identify unique *k-mers*, towards reducing the computational requirements for k-mer mapping as RDBMS functions, while improving the whole execution time. We present possible index structures and discuss their performance through time and I/O measurements. More specifically, we propose a novel index that takes advantage of consecutive *k-mers* properties, reducing the number of disk access and, consequently, the execution time. For a sugarcane genome dataset, our experimental results show that our approach achieves very good performances.

2   Related Work

Several techniques have been proposed to reduce the memory footprint during the assembly process [9] using a variety of data structures like suffix arrays, Bloom filters, FM indexes, and succinct data structures, focused directly or not on the K-mer Mapping process. We may divide these approaches into two strands: some of them try to reduce the amount of data, while others try to increase the resources for the same amount of data (partitioning or distributing).

On the one hand, they try to reduce the data (by removal or sampling) until the assembly can be executed with the available resources, but it may affect the final quality of assembly. On the other hand, we have greedy methods since they allocate more computational resources (*e.g.*, cloud structures) until they can sufficiently generate the assembly. Two approaches focused on reducing the K-mer Mapping memory footprint are presented in [12] and [10].

Minimum substring partition (MSP) [12] showed an intelligent solution based on breaking the short reads into multiple small disjoint partitions so that each partition could be loaded into memory, processed individually and later merged with others to build a *de Bruijn* graph. On the other hand, [10] presents an efficient parallel approach for constructing large *de Bruijn* graphs, which is also extensible to the out-of-core model. After generating all canonical edges, they detect duplicate *k-mers* by sorting all the edges using radix sort algorithms.

Additionally, there are tools developed for *k-mer* counting processes that are not exclusively useful for genome assembly. Jellyfish [14] is based on exploiting shared memory parallel computing, implementing a multi-thread algorithm over a lock-free hash table and a compact memory hash entry for counting *k-mers* up to 31 bases in length. The authors also briefly discussed the impact of varying the *k* value on computation time, showing the contribution of I/O.

DSK (Disk Streaming) [16] and KMC (K-mer Counter) [6] are two solutions for *k-mer* counting based on partition data processing. The reads are partitioned and each partition is separately loaded in the memory. We obtain the final result by merging independent solutions. DSK uses a temporary hash table to process *k-mers*. KMC also exploits CPU and I/O parallelism and uses the concept of prefixes to reduce data, and radix sort algorithm to detect duplicate *k-mers*.

Assembly is highly sensitive to data and the k value. The success of memory reduction techniques is only checked when the assembling is achieved. Memory requirements are higher for complex organisms (e.g. plants) that are nearly 100 times larger than mammalian genomes, with a higher degree of repetition [3,19].

3   Index Structures

A *de Bruijn* graph $G = \{V, E\}$ is built by creating a node for every unique *k-mer* in the U universe, over a set of sequences S, with the alphabet $\Sigma = \{A, T, C, G\}$. An edge between two nodes is created whenever their corresponding *k-mers* are adjacent in at least one short read. The K-mer Mapping is a process that identifies unique *k-mers*, and map all duplicate *k-mers* into the unique corresponding node in the *de Bruijn* graph, given a specific k. Therefore, an exact match over k length strings is necessary to identify duplicate *k-mers*.

In such a problem, the use of a data structure for storage and quick access is essential. The unique *k-mers* set is the only one that we need to store into a high-speed data structure, while the duplicate *k-mers* set only needs to be mapped over the former. Therefore, the number of unique *k-mers* determines the size of the structure for a duplicates search and defines the size of the nodes set of the *de Bruijn* graph. It should be noted that no deletions, nor updates occur.

The *k-mer* is a string whose length is $k, 1 < k < m$, where m a short read length. The *k-mers* are generated sequentially, using a sliding window with k width and step 1. Given s and $s[i, j]$, the substring of s between the ith and j^{th} (both inclusive) elements, the *k-mer* generation consist of breaking s into consecutive and overlapped $(m - k + 1)k$-*mers*, written as $s[1, k], s[2, k + 1], ..., s[m - k + 1, m]$. The *k-mers* $s[i, k + i - 1], s[i + 1, k + i]$ are adjacent in s and share $k - 1$ characters. A *k-mer* $s[i, k + i - 1]$ extracted from s_r is represented as $s_{r,i}$.

Then, a K-mer Mapping is a function that, given the set $U = \emptyset$ of unique *k-mers*, a set $M = \emptyset$ of duplicate *k-mer* maps, and a *map()* function that generates a map for a specific *k-mer*, for all *k-mer* generated $s_{r,i}$ executes *if* $s_{r,i} \notin U$ *then* $U = U \bigcup s_{r,i}$ *else* $M = M \bigcup map(s_{r,i})$.

The total number of *k-mers* present in one read is equal to $m - k + 1$, while the total number of *k-mers* present in n reads is $(m - k + 1) \times n$. The unique *k-mer* space for k value is 4^k, which is equal to 2^{2k}, an exponential function.

We define four entities and corresponding relations:

- *sequence*: {sequence string (m length and $\Sigma = \{'A','T','C','G'\}$),
sequence identification (number)}$\rightarrow$ **sequences relation**

- *presequence*: {preprocessed sequence string (m length and $\Sigma = \{'A','T','C','G','N'\}$),
sequence identification (number)}$\rightarrow$ **presequences relation**

- *unique k-mer*: {string (length k and $\Sigma = \{'A','T','C','G'\}$), read identification (number), position into the read (number)}$\rightarrow$ **identification relation**

- map: {read id (number), position in the read (number), read overlapped id (number), length of overlapped chain(number)}$\rightarrow$ **roadmap relation**

Each map is structured as (***read_id, position, read_id_over, length_cod***) where ***read_id*** is the read identifier to which belongs the *k-mer*, ***position*** is the relative position of the *k-mer* in the read, ***read_id_over*** is the read overlapped identification, and ***length_cod*** is equal to the *overlap string length - k + 1*.

We have implemented five main functions. The first two load the reads from 'fastq' or 'fasta' files into the **sequences** table for later, preprocessing them, replacing $'N'$ character for a valid nitrogenous base character, and loading it into **presequences** table. The third function executes the K-mer Mapping over sequences in **presequences** relation, searching for duplicate *k-mers* over **identification** relation, inserting unique *k-mers*, if needed, into **identification** relation, or mapping a duplicate *k-mer* into **roadmap** relation. Another function generates the maps into a file similar to the Velvet Roadmaps file. All of these functions are carried out by a main function, which also executes the transaction checkpoints to persist the data during the whole process.

The function that executes the K-mer Mapping was modeled based on (test case) the strategy implemented by Velvet, since it maps consecutive overlapped strings that contain more than one *k-mer* instead of mapping a simple *k-mer* each time that is duplicated. This strategy contributes to starting the next steps with a more compact *de Bruijn* graph, which is more efficient.

A sketch of our algorithm is outlined in Algorithm 1. It was implemented in PL/pgSQL and executed using PostgreSQL 9.5. The K-mer Mapping process visits each *k-mer* in **presequences** relation after preprocess sequences, and defines if it has a duplicate *k-mers* or not. For that, each *k-mer* extracted from each sequence must be searched in the **identification** relation, asking whether or not it already exists, *i.e.* does it have duplicates or not? At that point the index fulfills its function. If a *k-mer* K is found, it is checked if the left adjacent *k-mer* K_p, also has an overlap in the same read as the current *k-mer* K, as well as if it is the left adjacent *k-mer* of K in the overlapped read. The corresponding actions and control variables used for this process were encapsulated in the *isTheLeftAdjInOverlappedRead()* method in the algorithm sketch. If K_p meets the previous condition, the current *k-mer* K - with its info encapsulated

Algorithm 1. K-mer Mapping as RDBMS function

```
1:  for each sequenceRow IN SELECT * FROM presequences do
2:      initializes map: A = NULL
3:      initializes previous k-mer: Kp = NULL
4:      for readNucleotideIndex IN 1..(m − k + 1)  do
5:          k-mer: K = substr(sequenceRow.read, readNucleotideIndex, k);
6:          SELECT * INTO kmerRow FROM identification WHERE kmer = K;
7:          if kmerRow != NULL then
8:              if isTheLeftAdjInOverlappedRead(Kp, A, kmerRow) then
9:                  A = add(A, kmerRow);
10:             else
11:                 if A! = NULL then
12:                     INSERT INTO roadmap (read, overlappedRead, posInRead,
13:                     posInOverlappedRead,       numbOfContinuosKmerOverlapped,
        annotationOrder)
14:                     VALUES
15:                     (sequenceRow.id,            A.referenceSeqId,            A.pos,
        A.referenceSeqStartPos, A.finish, A.orderNumb );
16:                     A = NULL;
17:                 A = add(A, kmerRow));
18:         else
19:             INSERT INTO identification (kmer, read, pos) VALUES (kmer-
        String,sequenceRow.id,writeNucleotideIndex);
20:             if A! = NULL then
21:                 INSERT INTO roadmap (read, overlappedRead, posInRead,
22:                 posInOverlappedRead, numbOfContinuosKmerOverlapped, annota-
        tionOrder)
23:                 VALUES
24:                 (sequenceRow.id, A.referenceSeqId, A.pos, A.referenceSeqStartPos,
        A.finish, A.orderNumb );
25:                 A = NULL;
26:         Kp = K;
```

in kmerRow -, is included into the same map A following a process that tries to extend the overlapping region as much as possible. The update of map variables was encapsulated in the function add(). The map A is persisted into roadmap relation when no more consecutive k-mers are overlapped in a continuous chain, and a new map is created including the current k-mer. If K is not found into the identification relation, it is inserted with associated info, and map pending to be persisted (if there is any) is inserted into roadmap relation.

Only two kinds of operations are executed over the identification relation, searches and insertions. The number of searches $((m-k+1) \times n)$ and the number of insertions $(O(4^k))$ significantly influences the execution time.

Index Based on Minimum Substring

In [5] the use of two not clustered index configurations for *k-mers* have been studied in the context of *k-mer* Mapping: either tree-based indexes or a hash-based indexes [15]: (i) B+-tree as an index over the *k-mer* string (without nulls and uniqueness constraints), and; (ii) Hash index over *k-mer* string.

Both indexes were applied over the `kmer` attribute into the `identification` relation. The hash index over *k-mer* string stores data entries as *<key, rid>* pair similar to B+-tree index. Figure 2(a) has a representation of the index structure for the process of indexing S_r. For the first four *k-mers*, a hash function is applied to define the corresponding bucket. Each entry in the bucket for *k-mer* key points to the *k-mer* data record, which contains the value for each *k-mer*:{*k-mer* (string), *read identification* (number), *position into the read* (number)}. Nothing guarantees that adjacent *k-mers* share the same buckets or neighboring buckets.

However, a good K-mer Mapping index should be able to take advantage of the overlapped $k - 1$ length string shared by adjacent *k-mers*, to minimize the number of random I/O operations. Thus, a disk structure that keeps adjacent *k-mers* physically clustered is needed, since most I/O activity is generated searching *k-mers*, due to the high total number of *k-mers* $((m - k + 1) \times n)$ that must be searched, and the fact that the adjacent *k-mer* given one can be previously known with a probability of 1/4 (assuming uniform distribution).

An index structure capable to keep adjacent *k-mers* physically clustered must accomplish 3 properties: the index must be unique for an adjacent group, the index must be unique for each adjacent *k-mer* inside the adjacent group, and also the index must not depend on a particular position into the *k-mers* string.

Two adjacent *k-mers* in s, $s[i, k+i-1]$ and $s[i+1, k+i]$ share the substring $[i+1, k+i-1]$ with length $k-1$, while three adjacent *k-mers*, including the third $s[i+2, k+i+1]$ share the substring $[i+2, k+i-1]$ with length $k-2$, and so on (see Fig. 1). The existence of this overlapped substring for some number of adjacent *k-mers* suggests that it may be used to build an index capable of maintaining those adjacent *k-mers* physically close.

Fig. 1. K-mer generation. Adjacent *k-mers* overlapped substring

The minimum substring length-p (given $p \leq k - 1$) presents the properties already cited. Given a string s, a length-p substring r of s is called the minimum p-substring of s if $\forall s'$, s' is a length-p substring of s, s.t., $r \leq s'$ ($\leq$ defined by lexicographical order). The minimum p-substring of s is written as $min_p(s)$.

Given $s[i, k+i-1]$, the $min_p(s[i, k+i-1])$ is unique for $s[i, k+i-1]$, and by definition it is not tied to a specific position inside the *k-mer*. Since adjacent *k-mers* share *k-1* elements, it is highly likely that a chain of ℓ adjacent *k-mers* share the same $min_p(s_{r,i})$ while $p \leq k - \ell - 1$.

The minimum substring concept is also used over *k-mers* [22] to compact partitions. Using this partitioning technique, the authors in [12] implemented a partitioning, mapping and merging algorithm that was successfully applied to *de Bruijn* graph construction with very small memory footprint.

In this study, we introduce the use of a minimum substring as an index key for exact *k-mer* matches in an RDBMS environment for K-mer Mapping. Since it is unknown whether there exists a hash function capable of mapping two adjacent *k-mers* to the same bucket, we opted for using hashing over the $min_p(s_{r,i})$ as a strategy for mapping consecutive *k-mers* in the same bucket, thus, physically adjacent. With this approach it is more likely that two adjacent *k-mer* would be distributed to the same buckets (Fig. 2(b)), contrary to using a hash index over *k-mer* string (Fig. 2(a)). As a result, there is a decrease in the random I/O operations when the index does not fit in the main memory.

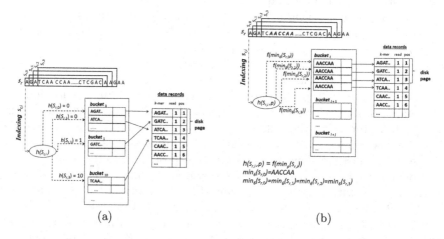

(a)   (b)

Fig. 2. (a) Hash index over *k-mer* string. (b) K-mer Mapping using a hash index over *k-mer* *p*-minimum substring.

Figure 2(b) shows that hashing the minimum *p*-substring (to illustrate we used $p = 6$) instead of *k-mer* itself, a group of *k-mers* was indexed in the same bucket. The index stores data entries as *<key, rid-list>* pair, where *rid-list* is a list of record ids of data records with search *key* value is key. In that case, the *key* is the *p*-minimum substring that points to data records with *k-mers* that share the same *p*-minimum substring. The figure is a representation of the structure of the index, for indexing the read *s*. As we can see for the first four *k-mers*, the hash function is applied $h(s_{r,i}, p) = f(min_p(s_{r,i}))$ to define the corresponding bucket, which is the same for all *k-mers* here since the four *k-mers* share the same $min_6(s_{r,i})$. This means that consecutive adjacent *k-mers* share the same index page or close ones, decreasing the I/O search overhead when the index does not entirely reside into the main memory. This approach increases the likelihood of the index pages for the current *k-mer* searched be already at the buffer. Each entry in the bucket for *k-mer* key points to the *k-mer* data record which contain the value for each *k-mer*:{*k-mer* (string), *read identifier* (number),*position into the read* (number), $min_p(k - mer)$ (string), *start position of* $min_p(k - mer)$ (number)}. The start position of $min_p(k - mer)$ relative to *k-mer* is used to compress whenever the *p*-minimum substring is not repeated into the *k-mer*.

4   Results and Discussion

We executed some tests in a CPU Core i7-3770 3.40 GHz, 8 GB RAM, and 1 TB of HDD machine, considering Ubuntu 14.0.4 LTS Linux distribution. The experiments ran over a sugarcane DNA with 2 million reads. The read length is 100, and we have used a sufficiently large *k* variation, over 39 to 71.

The values showed in Fig. 3 from results in [5] represent the K-mer Mapping runtime results varying *k* value for both indexes: B+-tree primary key over *k-mer* string and hash over *k-mer* string over 2 millon reads of sequences. Those results show that while *k* increases, the execution time decreases. This is because the total number of *k-mers* decreases.

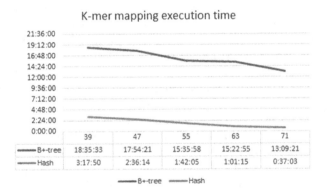

Fig. 3. K-mer Mapping execution time.

4.1   The Impact of *k* Value in the Processing Time

Biologists researchers generate assemblies using a range of *k* values over the same sequences dataset. After that, they select the best assembly. We are interested in studying the relationship between the runtime and *k* value for K-mer Mapping executions over the same sequences dataset. We could explore many interesting questions in this context, such as:

– As the total *k-mer* number for execution *A* is a half of the *k-mer* used for *B* execution, would it mean that *A* runtime would be a half of *B*?

- If we have execution A and B, with k_a and k_b the respectively k values, given $k_a > k_b$, will the runtime of A be higher than the runtime of B?
- If we run A and B, k_a and k_b the corresponding k values, $k_a < k_b$. Can we assume that B runtime is an upper bound for A runtime.

While k increases, for the same number of sequences, the number of unique k-mers decreases, but the percentage of unique k-mers increases over the total number of k-mers. Also, the total number of k-mers decreases at a ratio $r = 1 - (i - j)/(m - j + 1)$, where $i, j \in \mathcal{L}$ and $i > j$.

Figure 3 shows the runtime variation while k value increased for both indexes. Both indexes had the same behavior respect k, while k increases the execution time decreases. The principal cause of that is the total k-mers number decrease at ratio r. Effectively 64.000.000 less searches were executed for $k = 71$ compared with the execution with $k = 39$. Moreover, 20.215.018 unique k-mers less were inserted into the table that contains unique elements, decreasing the size of the index. That behavior is an evidence of the execution time is highly depend on the number of total k-mers.

Tables 1 and 2 show the proportions K-mer Mapping runtime using B+-tree primary key and hash indexes, respectively. In both cases, the execution time is not linear with respect to the total number of k-mers. However, for B+-tree, all execution times proportion were higher than r, while for hash index were less than r. Hence, given two executions A and B such that the total number of k-mer for A is a half of the used in B, the runtime expected for A using the B+-tree primary key index is higher than a half runtime of B, while for hash index is less than a half runtime of B. For example, the total number of k-mer for $k = 71$ is 48% of the total k-mer number for $k = 39$, but the execution time using B+-tree primary key for $k = 71$ represented 71% of the execution time for $k = 39$, while for hash index the execution time percentage was 19%.

The runtime proportion ($tr = $ runtime for $k = i$/runtime for $k = j$ with $i < j$) appears to be higher while increases the number of processing reads. Figure 4 shows the execution time for different k values as the number of processed reads increases, given $k \in \{39, 47, 55, 63, 71\}$. For the first 600,000 reads, tr (for any value of i and j, $i < j$) had small values (the colored lines on the chart are very close) for both indexes. For the next 600,000 reads, tr becomes greater. For B+-tree index, tr tends to remain constant given two pairs (i, j) while the reads processed increase, *i.e.*, the lines tends to be parallel. On the contrary, using a hash index, tr becomes more significant as more reads are processed.

Evaluation of p-Subtring Index. To evaluate our approach of using the hash over p-minimum substring index, we execute the same experiments but in different sequence datasets.

For 3 and 5 million reads datasets, the use of the index over $min_{36}(k\text{-}mer)$ brought significant gains in runtime (see Table 3). In these cases, the K-mer Mapping obtained the best runtimes using the hash over 36-minimum substring index for all k, with a considerable difference observed in a "Saved time" column. The saved times by using a hash over p-minimum substring index are influenced

Table 1. K-mer Mapping runtime for B+-tree index

k	Number of total k-$mers$	i	j	r	B+-tree index		
					Run-time	tr	tr/r
Comparing each $i = j + 8$							
39	124.000.000	0	0	0	18:35:33	0	0
47	108.000.000	47	39	0,8710	17:54:21	0,96	1,11
55	92.000.000	55	47	0,8519	15:35:58	0,87	1,02
63	76.000.000	63	55	0,8261	15:22:55	0,99	1,19
71	60.000.000	71	63	0,7895	13:09:21	0,86	1,08
Comparing each i with $j = 39$, the less k value							
39	124.000.000	0	0	0	18:35:33	0	0
47	108.000.000	47	39	0,8710	17:54:21	0,96	1,11
55	92.000.000	55	39	0,7419	15:35:58	0,84	1,13
63	76.000.000	63	39	0,6129	15:22:55	0,83	1,35
71	60.000.000	71	39	0,4839	13:09:21	0,71	1,46

Table 2. K-mer Mapping runtime for Hash index

k	Number of total k-$mers$	i	j	r	Hash index		
					Run-time	tr	tr/r
Comparing each $i = j + 8$							
39	124.000.000	0	0	0	3:17:50	0	0
47	108.000.000	47	39	0,8710	2:36:14	0,79	0,82
55	92.000.000	55	47	0,8519	1:42:05	0,65	0,75
63	76.000.000	63	55	0,8261	1:01:15	0,60	0,61
71	60.000.000	71	63	0,7895	0:37:03	0,60	0,71
Comparing each i with $j = 39$, the less k value							
39	124.000.000	0	0	0	3:17:50	0	0
47	108.000.000	47	39	0,8710	2:36:14	0,79	0,82
55	92.000.000	55	39	0,7419	1:42:05	0,52	0,62
63	76.000.000	63	39	0,6129	1:01:15	0,31	0,37
71	60.000.000	71	39	0,4839	0:37:03	0,19	0,26

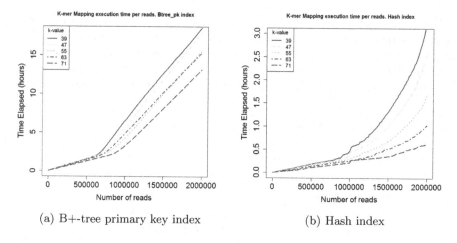

(a) B+-tree primary key index   (b) Hash index

Fig. 4. K-mer Mapping runtime for different k value per number of reads processed.

Table 3. K-mer Mapping runtime for 3 and 5 million reads

k	K-mer Mapping runtime					
	Over 3 million reads			Over 5 million reads		
	Hash index over k-mer	Hash index over $min_{36}(k\text{-}mer)$	Saved time	Hash index over k-mer	Hash index over $min_{36}(k\text{-}mer)$	Saved time
39	17:10:59	13:08:52	4:02:07	48:30:54	28:31:13	19:59:41
47	15:16:47	7:27:59	7:48:48	44:56:39	18:54:39	26:02:00
55	13:14:15	6:51:07	6:23:08	39:29:09	16:08:49	23:20:20
63	8:47:21	5:38:25	3:08:56	34:01:25	13:46:08	20:15:17
71	6:09:50	4:08:28	2:01:22	23:03:51	11:04:22	11:59:29

by the number of searches and insertions executed during the K-mer Mapping, which were larger than their counterparts for 2 million reads. To support these claims, we can observe that for $k = 47, 55, 63$ *and* 71, there is a trend towards a decrease in the saved times, while k increases, as the number of searches and insertions decrease too while k increases. However, the time saved for $k = 39$ does not follow this trend, *i.e.*, does not have the expected value. This is due to the difference between k, and p was rather small, leaving only three bases as the maximum difference between two adjacent k-*mer* to take advantage of an index. Also, for the execution with $k = 71$, our goal would be to improve the saved time. Particularly, for the 5 million reads dataset, where the number of searches and insertions is greater than their 3 million reads dataset counterparts. The saved time difference value is more evident. For execution with $k = 71$, k is almost the double of p, a large difference, which may not allow the index to be exploited to the maximum. To illustrate the index performance, we selected $p = 36$ for all experiments, since $p < k$ for all cases, but an adjusted p value for each k would be beneficial. The fact that p values influence the K-mer Mapping execution time is clear, but this analysis is out of the scope of this work.

The execution times, considering a hash over p-minimum substring index, represent less than the 77% and 60% (for 3 and 5 million reads processed correspondingly) of the execution times using a hash over k-*mer* string index. We even managed to save up 51% and 60% of the execution times, with $k = 47$ in the 3 million reads dataset, and $k = \{55, 63\}$ for the 5 million reads dataset, respectively, demonstrating the improvement in efficiency brought by the use of p-minimum substring index.

5 Conclusions

In this work, we present a study of an index configuration approach to K-mer Mapping into RDBMS, as a strategy to avoid the high main memory consumption for plant genome *de novo* assembling.

We propose a novel index approach based on hashing the p-minimum substring of k-*mers* as a strategy to increase the number of buffer hits for adjacent k-*mer*. Consequently, it reduces the I/O operations. The results show better performance for the hash index over p-minimum substring of the k-mer, saving 48% of time processing 5 million reads for variable k.

Despite the improvements in runtime brought by the new index approach, the time resulting from our experiments could still be considered high. We plan to introduce an optimization based on the precision of the value of p and to apply other compression strategies used in homologous problems, such as counting k-*mers* to reduce the execution time of K-mer Mapping in RDBMS.

References

1. Bradnam, K.R., Fass, J.N., et al.: Assemblathon 2: evaluating de novo methods of genome assembly in three vertebrate species. GigaScience **2**(1), 1–31 (2013)
2. Butler, J., et al.: ALLPATHS: de novo assembly of whole-genome shotgun microreads. Genome Res. **18**(5), 810–820 (2008)
3. Claros, M.G., Bautista, R., Guerrero-Fernández, D., Benzerki, H., Seoane, P., Fernández-Pozo, N.: Why assembling plant genome sequences is so challenging. Biology **1**(2), 439 (2012)
4. Cook, J.J., Zilles, C.: Characterizing and optimizing the memory footprint of de novo short read DNA sequence assembly. In: Performance Analysis of Systems and Software, ISPASS, pp. 143–152, April 2009
5. de Armas, E.M., Haeusler, E.H., Lifschitz, S., de Holanda, M.T., da Silva, W.M.C., Ferreira, P.C.G.: K-mer Mapping and de Bruijn graphs: the case for velvet fragment assembly. In: Proceedings IEEE International Conference on Bioinformatics and Biomedicine (BIBM), pp. 882–889 (2016)
6. Deorowicz, S., Debudaj-Grabysz, A., Grabowski, S.: Disk-based k-mer counting on a PC. BMC Bioinform. **14**(1), 160 (2013)
7. Earl, D., Bradnam, K., et al.: Assemblathon 1: a competitive assessment of de novo short read assembly methods. Genome Res. **21**(12), 2224–2241 (2011)
8. El-Metwally, S., Hamza, T., Zakaria, M., Helmy, M.: Next-generation sequence assembly: four stages of data processing and computational challenges. PLoS Comput. Biol. **9**(12), 1–19 (2013)

9. Kleftogiannis, D., Kalnis, P., Bajic, V.B.: Comparing memory-efficient genome assemblers on stand-alone and cloud infrastructures. PLoS ONE **8**(9) (2013)
10. Kundeti, V., Rajasekaran, S., Dinh, H.: Efficient parallel and out of core algorithms for constructing large bi-directed de Bruijn graphs. ArXiv (2010)
11. Li, R., Zhu, H., et al.: De novo assembly of human genomes with massively parallel short read sequencing. Genome Res. **20**, 265–272 (2009)
12. Li, Y., Kamousi, P., Han, F., Yang, S., Yan, X., Suri, S.: Memory efficient minimum substring partitioning. PVLDB **6**(3), 169–180 (2013)
13. Luo, R., et al.: SOAPdenovo2: an empirically improved memory-efficient short-read de novo assembler. GigaScience **1**(1), 1–6 (2012)
14. Marcais, G., Kingsford, C.: A fast, lock-free approach for efficient parallel counting of occurrences of k-mers. Bioinformatics **27**(6), 764–770 (2011)
15. Ramakrishnan, R., Gehrke, J.: Database Management Systems, 3rd edn. McGraw-Hill Inc, New York, NY, USA (2003)
16. Rizk, G., Lavenier, D., Chikhi, R.: DSK: k-mer counting with very low memory usage. Bioinformatics **29**(5), 652–653 (2013)
17. Salzberg, S.L., et al.: GAGE: a critical evaluation of genome assemblies and assembly algorithms. Genome Res. **22**(3), 557–567 (2012)
18. Schatz, M.C., Delcher, A.L., Salzberg, S.L.: Assembly of large genomes using second-generation sequencing. Genome Res. **20**(9), 1165–1173 (2010)
19. Schatz, M.C., Witkowski, J., McCombie, W.R.: Current challenges in de novo plant genome sequencing and assembly. Genome Biol. **13**(4), 1–7 (2012)
20. Simpson, J.T., Wong, K., Jackman, S.D., Schein, J.E., Jones, S.J., Birol, I.: ABySS: a parallel assembler for short read sequence data. Genome Res. **19**(6), 1117–1123 (2009)
21. Zerbino, D.R., Birney, E.: Velvet: algorithms for de novo short read assembly using de bruijn graphs. Genome Res. **18**(5), 821–829 (2008)

A Clustering Approach to Identify Candidates to Housekeeping Genes Based on RNA-seq Data

Edian F. Franco[1,2,6], Dener Maués[2], Ronnie Alves[3], Luis Guimarães[1],
Vasco Azevedo[4], Artur Silva[1], Preetam Ghosh[5], Jefferson Morais[2],
and Rommel T. J. Ramos[1,2(✉)]

[1] Institute of Biological Sciences, Laboratory of Biological Engineering,
Federal University of Para, Belem, Para, Brazil
edianfranklin@gmail.com, rommelramos@gmail.com
[2] Department of Computer Science,
Computer Science Postgraduate Program (PPGCC), Federal University of Para,
Belem, Para, Brazil
jeffersonmorais@gmail.com
[3] Vale Technology Institute, Belem, Para, Brazil
alvesrco@gmail.com
[4] Institute of Biological Sciences,
Federal University of Minas Gerais-UFMG, Belo Horizonte, Minas Gerais, Brazil
vascoariston@gmail.com
[5] Department of Computer Science, Virginia Commonwealth University,
Richmond, VA, USA
pghosh@vcu.edu
[6] Basic and Environment Science Department, Instituto Tecnológico de Santo
Domingo (INTEC), Santo Domingo, Dominican Republic

Abstract. Housekeeping genes (HKGs), are essential for gene expression based studies performed through Reverse Transcriptase-polymerase Chain Reaction (RT-qPCR). These genes are related with the basic cellular processes that are essential for cell maintenance, survival and function. Thus, HKGs should be expressed in all cells of an organism regardless of the tissue type, cell state or cell condition. High-throughput technologies, including RNA sequencing (RNA-seq), are used to study and identify these types of genes. RNA-seq is a high-throughput method that allows the measurement of gene expression profiles in a target tissue or an isolated cell. Moreover, machine learning methods are routinely applied in different genomics related areas to enable the interpretation of large datasets, including those related to gene expression. This study reports a new machine learning based approach to identify candidate HKGs *in silico* from RNA-seq gene expression data. The approach enabled the identification of stable HKGs candidates in RNA-seq data from *Corynebacterium pseudotuberculosis*. These genes showed stable expression under different stress conditions as well as low variation index and fold changes. Furthermore, some of these genes were already reported in the literature as HKGs or HKGs candidates for the same

© Springer Nature Switzerland AG 2020
L. Kowada and D. de Oliveira (Eds.): BSB 2019, LNBI 11347, pp. 83–95, 2020.
https://doi.org/10.1007/978-3-030-46417-2_8

or other bacterial organisms, which reinforced the accuracy of the proposed method. We present a novel approach based on K-means algorithm, internal metrics and machine learning methods that can identify stable housekeeping genes from gene expression data with high accuracy and efficiency.

Keywords: Housekeeping genes · Machine learning · Clustering

1   Introduction

Housekeeping Genes (HKG) are constitutive genes required for the maintenance of basic cellular functions. Thus, HKGs are expressed in all cells of an organism under both normal and pathophysiological conditions [13], these genes are important to understand the transcriptional processes and cell cycle life [24]. Hence, the expression of HKGs is used as reference point in the expression levels of other genes in analyzing gene expression datasets [11].

The gene expression data is essential to understand many biological processes in different organisms in relation to its environment. RNA-Seq is an important technique frequently used in recording gene expression data that allows for capturing an accurate picture of the molecular processes within the organisms and understanding the interaction between genes or other genomic elements [8,27,37,39]. The HKGs identification is no a trivial process, several gene expression data analysis methods are available to identify HKGs, including machine learning (ML) methods [26], which enable the collection and identification of valid, new, potentially usable and understandable patterns and knowledge based on a dataset, which can lead to the detection and identification of possible HKGs candidates [14,23].

Within machine learning methods, clustering algorithms are based on unsupervised learning and are used to cluster objects based on the intrinsic information contained in the data and their relationships [18]. Clustering algorithms seek groups that are a) compact, with the members of each cluster as close as possible to one another, and b) separated, in which the clusters are as far apart as possible [20]. These methods may be applied for the identification of expression patterns of gene groups, including HKGs [22].

Clustering algorithms use different methods or techniques to discover clusters. The most popular methods applied to the bioinformatics domain are the partitioned, hierarchical, density-based, and grid-based methods [9,18]. The partitional clustering algorithms as K-means are the most used clustering methods, that divided the data into several subset, with each cluster optimized the clusters criterion and minimizated the sum of squared distance from the mean within each cluster [1,2]. Internal metrics are used to evaluate the clustering methods and seek the set of groups that best adapt to the natural partitions [31]. These metrics evaluate cluster compactness, separation and robustness using information intrinsic to the data [5].

This study reports an unsupervised machine learning based approach to identify housekeeping genes candidates based on RNA-seq data. The analysis is supported by intrinsic evaluation metrics of clustering to evaluate and validate our method. We employed data from expression studies of *Corynebacterium pseudotuberculosis* [28] that were exposed to different levels of stress to test our method.

2   Related Work

Some studies related to HKGs identification have been conducted using ML approaches with conventional classifiers (e.g., neural networks and support vector machine (SVM)). Some approaches predict housekeeping genes based on Fourier transform analysis of time-series gene expression data in combination with SVM; that method has been used to identify 510 HKGs in human genomic data [12]. Studies have classified HKGs based on physical and functional characteristics using exon length and chromatin compactness measurements as properties with the Naive Bayes classification algorithm to identify new HKGs candidates in humans [10]. Other studies used a specialized database to identify HKGs using SVM in RNA-seq datasets, this method found 1161 HKGs in the database [30]. These studies used eukaryotic data, which demonstrates the viability of identifying these genes in other organisms, including prokaryotes, using ML methods.

3   Methods

3.1   Proposed Pipeline for HKG Identification

The proposed approach consists of three stages (Data processing, clustering and HKGs candidates identification) (Fig. 1):

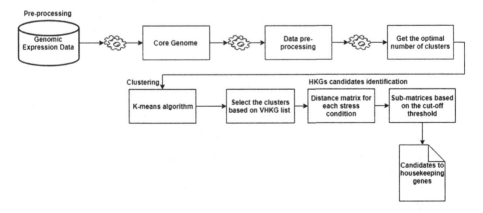

Fig. 1. Flowchart of the approach used to identify candidate housekeeping gene candidates using expression data

3.2   Data Processing

RNA-seq Data. The RNA-seq data used in this study was gathered from the study by Pinto and colleagues (2014), wherein *Corynebacterium pseudotuberculosis* strains CP1002 and CP258 were subjected to different stress levels (acid, osmotic and heat stress) using the RNA-seq method in the SOLiD System sequencing platform [28].

The RNA-seq data was processed through a reference-based approach using the CLC Genomics Workbench software and considering alignments with at least 70% identity in 80% of the read lengths using the following *Corynebacterium pseudotuberculosis* strains: CP258, isolated from a sheep (CP003540.2) [36], and CP1002, isolated from a horse (CP001809.2) [35].

For the CP258 and CP1002 genomes, the genes belonging to the core genome was obtained using the Pan-Genome Analysis Pipeline (PGAP) software [40] which enabled us to identify the set of genes shared between the *Corynebacterium pseudotuberculosis* genomes which were then designated as possible HKGs.

Thus, 38 *Corynebacterium pseudotuberculosis* strains were used (12 from horses and 26 from sheep) with 90% coverage, 90% identity and an e-value of 1E−05 as parameters. The strains were divided into two groups according to the biovars (equi and ovis).

Data Pre-processing. The raw RNA-Seq data were normalized by reads per kilobase of exon model per million mapped reads (RPKM). The RPKM values are transformed into ($\log_2 +1$), thereby reducing data asymmetry and the outlier effects on the dataset [25]. Moreover, the data-sets was processed to find and remove the extreme values and outlier expression values to improve the accuracy of the clustering algorithm following the methodology shown in [27].

Determining the Optimal Number of Clusters. All expression datasets were evaluated using the SD validity (SDbw) index and Dunn index as defined below; these indices can individually predict the optimal number of clusters in each dataset. Alongside, some other indices were also evaluated, such as, Silhoutte index and connectivity, to selected the better index to according the data-sets.

SDbw Index. This validation index definition is based on the criteria of compactness and separation between clusters [6,16]. The index is computed by the Eq. 1:

$$SDbw(q) = Scat(q) + Density.bw(q) \qquad (1)$$

The first term, Scat(q), signifies the average compactness of q clusters (i.e., intra-cluster distance) and the second term $Density.bw(q)$, is the inter-cluster density. It evaluates the average density in the region among clusters in relation to the density of the clusters.

Dunn Index. The Dunn index was used to identify compact clusters with good separation. The index is defined as [3]:

$$
\min_{1 \leq i \leq c} \left\{ \min_{\substack{1 \leq j \leq c \\ j \neq i}} \left\{ \frac{\delta(X_i, X_j)}{\max\limits_{1 \leq j \leq c} \{\Delta(X_k)\}} \right\} \right\}
\tag{2}
$$

wherein $\delta(X_i, X_J)$ defines the distance between clusters X_i and X_j (intercluster distance), $\Delta(X_k)$ represents the intracluster distance of cluster X_k, and c is the cluster number in the partition [3].

The indices were implemented with a variation in the number of clusters in the 2–20 range for each of the selected algorithms. The data sets were evaluated using cluster internal metrics according to Eqs. 1 and 2.

The R suite [29] was used to implement this metric with the specialized packages NbClust [15], and clValid [4]. In the subsequent analysis, we used the average of the number of clusters from each evaluation metric to serve as the best number of clusters to generate the results.

3.3   Clustering

To identify HKGs candidates based on RNA-seq data, the clustering algorithm was used on the expression profiles of the genes to find similarities between the genes in the datasets [34]. The partitioning (k-means) algorithm was explored on the gene expression data; this algorithm is a distance and centroid based, and popular in exploring gene expression dataset [9,27].

We used the number of clusters (k), based on the optimal number of clusters identified in the previous step to perform the algorithm. The k-means algorithm was executed based on the Euclidean distance between the genes. The Weka data mining software [17] and R language were used to perform the clustering analysis [33].

3.4   HKGs Candidates Identification

List of Validated Housekeeping Genes (VHKGs). To select the housekeeping genes candidates list, we identified a set of HKGs genes described in the literature to use as references points on the clusters to select the new candidates based on their proximity with these genes. This generic list is formed of genes that were used in bacterial expression studies [22].

A total of 9 genes *(ftsZ, gap, gyrA, gyrB, recA, secA, rho, rpoA and rpoB)* identified in the literature as VHKGs were selected as the generic list for all the bacterial genomes used in the study. The selection was based on the criteria of [32], who identified a set of reference genes for bacterial studies through a literature review and then selected the genes validated by two or more RT-PCR studies.

After applying each clustering algorithm, we selected only those clusters where one or more VHKGs were present; this allows to reduce the search space and select the genes that have greater proximity to the VHKGs with respect to their Euclidean distance.

Distance Matrix Calculation for Each Expression Condition. The distance matrix of each gene of the cluster was calculated with respect to the VHKGs based on the Euclidean distance, which is a metric commonly used to assess dissimilarity. Thus, the Euclidean distance between points X and Y in a Euclidean p-space is calculated according to Eq. 3:

$$d_{euc}(X, Y) = \left(\sum_{j=1}^{p} (x_j - y_j)^2 \right)^{\frac{1}{2}} \tag{3}$$

The Euclidean distance was calculated for each dataset (i.e., strain) under each stress condition based on the gene expression values.

Based on the Euclidean distance matrices, a heuristic method was used to identify the genes closest to the VHKGs and also the genes whose gene expression levels did not vary significantly under the different stress conditions to enable the selection of candidates housekeepig genes

Creation of Sub-matrices Based on the Cut-Off Threshold. Sub-matrices were created to select the closest genes based on the Euclidean distance matrix using the first distance quartile ($q1$) as a cut-off threshold because this threshold included the elements close to the VHKGs. Then, the filter $y \leq q_1$ (y is any gene in the matrix) was applied if under a specific stress condition, the gene belongs to the sub-matrix. The distances of each gene present in the sub-matrices of the different conditions in relation to the VHKGs were next analyzed to identify the closest genes as possible HKGs candidates.

If y is an VHKGs with constant expression and little variation in all stress conditions $C_1, C_2, C_2 \ldots C_n$ and A is a gene with little variation in its expression level under all conditions and is close to y in relation to (q_1), then A is a possible candidate housekeeping gene for the strain under consideration in Fig. 2. The set of genes obtained using this approach was identified for each clustering algorithm.

Filtering and Validation of Possible Candidates to Housekeeping Genes. To filter the genes identified as possible HKGs candidates, a stability test was performed based on the coefficient of variation (CV), which was used as a statistical metric to compare the degree of variation between genes regardless of their mean expression levels [19], as defined by the Eq. 4:

$$CV = \frac{S}{X} \tag{4}$$

where S is the standard deviation of gene expression and X is the mean gene expression level under the different stress conditions. This method enabled the selection of genes with the lowest expression variation by setting $CV < 15$ as the threshold. Another metric used to evaluate candidates was the standard deviation (SD); we only selected genes with an $SD < 1$ [19].

These statistical metrics enable the identification of genes with stable expression levels, low coefficients of variation and low standard deviation between strains as candidate HKGs.

Approach for Accuracy Test. To test the accuracy of the approach, we selected the list of essential and conditional genes (where, conditional genes are the ones that have not been widely adopted by existing essential gene databases), using the *Mycobacterium tuberculosis*, a phylogenetically close organism of the *C. pseudotuberculosis*. The datasets were obtained from [7]. We performed a scoring test to check how many of these reported essential or conditional genes can be identified by our approach using the same set of 9 VHKGs as mentioned before, using the Eq. 5:

$$Accuracy = \frac{TGE + TGC + VHKG}{TGI} \tag{5}$$

where TGE is the number of essential genes identified, TGC is the number of conditional genes identified, $VHKG$ is the number of validated housekeeping genes and TGI is the total of genes identified by the approach. This test allows us to verify the ability of our method in identifying the possible candidate housekeeping. A high score obviously suggests that our candidate genes can be classified as an essential or conditional gene, and possess greater probabilities of being an actual housekeeping [21].

4   Results and Discussion

PGAP [40] was used to identify the genes of the core genome based on the pangenomic approach. We identified 1285 genes for strain Cp-258 belonging to the core genome for the set of 12 genomes of the biovar equi, 1191 of which (55% of total) showed gene expression in the strain. For the 26 genomes of biovar ovis, 1072 of the 1116 genes belonging to the core genome (54% of the total genome) were expressed in strain 1002.

The expression data were transformed into $log_2 + 1$, which allowed us to obtain a dataset closer to real values by suppressing outliers. Thus, we observed that clustering algorithms based on distance showed good performance in defining the border separating the clusters.

Outliers and extreme values were detected used the Interquartile range formula $[Q_1 - k(Q_3 - Q_1), Q_3 + k(Q_3 - Q_1)]$ where k is a constant and $IQR = Q_3 - Q_1$. The IQR was implemented used the InterQuartileRangeFilter from Weka [17]; this filter uses $k = 3$, to define the genes as an outlier and $K = 3 * 2$ to define an instance as extreme value, and the formula guarantees that at least 50% values are considered non-outliers. We got as final data sets: 1173 genes for Cp258 and 1109 genes for Cp1002 (Table 1).

Table 1. Data processing to obtain the final genome datasets

Genomes	Core genome genes	Genes of the core genome expressed in the genome	Genes with null values, outliers and extreme value	Final dataset
Cp-258	1285	1191	18	1173
Cp-1002	1114	1110	1	1109

The data-sets were evaluated using the indices: SDdw and Dunn using Nbclust [15] and Clvalid [4] packages in R [29] to identify the optimal number of clusters on the data-sets, these packages performed the k-means algorithm and applied the metrics to get the better division the datasets. We took the average of the result of both metrics as the optimal number of clusters using expression data for genes as packages input. To select the optimal number of clusters we implemented the metrics with the number between 2 to 20 clusters and Table 2 shows the results for each strain.

Table 2. Evaluation indices to get the optimal number of clusters for each strain.

Indexes	Cp-258	Cp-1002
SDbw	18	11
Dunn	9	17
Average	**13**	**14**

The K-means clustering algorithm was implemented based on the average number of cluster results from the cluster evaluation metrics using Weka and R. Figure 2 shows the graphical distribution of the clustering for each strain.

After clustering, the list of VHKGs for each of the genomes were adapted, based on the genes present in that specific genome. Table 3 shows the list of VHKGs adapted for each genome. This allowed selecting only the clusters where one or more VHKGs were present. For each of these clusters with VHKGs, we calculated the Euclidean distance to identify the genes that are closest to each VHKGs.

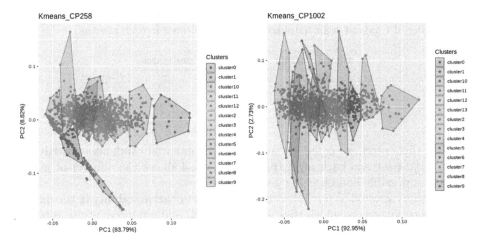

Fig. 2. Cluster size for each dataset and algorithm. The graphical distribution of the clusters in each data-set with the different algorithms applied. These points are projections of each genome on the first two principal components (PCs)

Table 3. Adapted validated housekeeping genes list. Adapted VHKGs list for each genome based on [32] where all the genes have been tested in three or more studies

	Adapted validated VHKGs genes list
Cp-258	gap, gyrA, gyrB, recA, rho, rpoA, rpoB and secA
Cp-1002	ftsZ, gap, gyrA, gyrB, recA, rho, rpoA, rpoB and secA

In order to select the cutoff of the Euclidean distance and the best variation coefficient to identify the possible candidates to HKGs, we experimented with different distance cut-offs and CVs. For this, we developed a cross-validation test based on the different combinations of using only the VHKGs, where we verify if using a specific cutoff these chosen VHKGs could find the other VHKGs in the data set. With this test we tried different combinations of cutoff (Euclidean distance: 1st quartile, median and mean and with CV: $<= 10, <= 15$ and $<= 20$; other metric tried was the M measure from [38]). With this test, we determined that the best cutoff for both metrics was the 1st quartile of the Euclidean distance (Q1), coefficient of variation less than 15 ($CV <= 15$) and the standard deviation less that 1 $SD < 1$. With these cutoffs, we could identify the genes that have greater proximity to the VHKGs and greater stability which is a desired characteristic for the housekeeping genes.

Table 4. Final candidate HKGs list using each genome strain. The table shows the final number of candidates for housekeeping genes selected in each genome

Genome strain	Final HKGs candidates
Cp-258	58
Cp-1002	76

With this cutoff, we could get a final list of potential candidates for HKGs, for each data set and each algorithm. The Table 4 shows the number of genes selected as possible housekeeping genes.

The genes selected as possible candidates shows relative stability in the different stress conditions to which they were subjected, and reinforces the possibility that they can be used as HKGs.

Results from the Accuracy Test. In order to develop this test, we excluded from the data sets of Cp-258 and Cp-1002, all the genes that were identified as hypothetical proteins, because they could not be classified in the categories we defined earlier for serving as reference (i.e., essential, conditional and non-essential). Table 5 shows the accuracy of the on the genomes strains in relation to the number of essential and conditional genes selected.

Table 5. Accuracy results. Result of the accuracy test of the different genomes strain, using the list of essential genes as reference

Genome strain	Total genes	Essential/conditional genes	Accuracy
Cp-258	38	25	65.78%
Cp-1002	58	41	70.68%

As can be seen in the results, the strain Cp-1002 obtained a better accuracy in the selection of a greater number of candidate genes.

Table 6. Final list of genes used for the accuracy test

Genome strain	Validated housekeeping genes	Essential genes	Conditional genes
Cp-258	gap, gyrA, gyrB, recA rho, rpoA, rpoB and secA	glmS, rplB, rplD, rpsC and secY	adk, aspB, ccdA, ctaC glgX, glpR, gltA, ispG pgk, pimB, ppc and tal
Cp-1002	ftsZ, gap, gyrA, gyrB, recA, rho, rpoA, rpoB and secA	cysS, cysS, dapA, efp, hisS, metG, rplC, rpsD, rpsE, trpS and ag84	aroK, asd, ctaC, gatB, glf, gpmA, ileS, leuS, nusB, ppa, ppc, pspA, ribD, rnc, rplR, rplT, rpoC, rpsI, rpsS, tatA, thrS and tpiA

Table 6 shows the final list of genes and their classification in the genomes that were used as a reference. We also identified the genes used as VHKGs in the approach and those that were identified as possible HKGs.

5 Conclusion

The proposed approach based on clustering algorithms and methods identified housekeeping genes from gene expression data generated by high-throughput sequencing platforms and could be adapted to other studies for establishing control genes for gene expression analyses.

Our method showed that it can adequately identify candidates with a high probability of being housekeeping genes, because of its stability in the expression at different levels of stress that were studied here.

The proposed methodology can be used with the set of genes of the core genome or with the whole genome, however, using the core genome allows improving the accuracy of the genes that are selected as candidates.

An important point is the selection of the genes that will be used as VHKGs. The result of the algorithms will be directly proportional to the accuracy of these genes.

We only tested our methodology and pipeline in prokaryote data, but our approach is also extendable for use in the eukaryotic genome. As future work, we will implement the eukaryotic version of the house keeping gene identification software.

References

1. Andritsos, P., et al.: Data clustering techniques. Rapport technique. University of Toronto. Department of Computer Science (2002)
2. Berkhin, P.: A survey of clustering data mining techniques. In: Kogan, J., Nicholas, C., Teboulle, M. (eds.) Grouping Multidimensional Data, pp. 25–71. Springer, Heidelberg (2006)
3. Bolshakova, N., Azuaje, F.: Cluster validation techniques for genome expression data. Sig. Process. **83**(4), 825–833 (2003). https://doi.org/10.1016/S0165-1684(02)00475-9
4. Brock, G., Pihur, V., Datta, S.: clValid: an R package for cluster validation. J. Stat. Softw. **25**, 1–32 (2008)
5. Brun, M., et al.: Model-based evaluation of clustering validation measures. Pattern Recogn. **40**(3), 807–824 (2007)
6. Charrad, M., Ghazzali, N., Boiteau, V., Niknafs, A.: NbClust: an R package for determining the relevant number of clusters in a data set. J. Stat. Softw., Art. **61**(6), 1–36 (2014). https://doi.org/10.18637/jss.v061.i06
7. Chen, W.H., Minguez, P., Lercher, M.J., Bork, P.: OGEE: an online gene essentiality database. Nucleic Acids Res. **40**(D1), D901–D906 (2011)
8. Ching, T., et al.: Opportunities and obstacles for deep learning in biology and medicine. J. R. Soc. Interface **15**(141), 20170387 (2018)

9. Dalton, L., Ballarin, V., Brun, M.: Clustering algorithms: on learning, validation, performance, and applications to genomics. Curr. Genomics **10**(6), 430–445 (2009). https://doi.org/10.2174/138920209789177601
10. De Ferrari, L., Aitken, S.: Mining housekeeping genes with a Naive Bayes classifier. BMC Genomics **7**(1), 277 (2006). https://doi.org/10.1186/1471-2164-7-277
11. Dheda, K., Huggett, J.F., Bustin, S.A., Johnson, M.A., Rook, G., Zumla, A.: Validation of housekeeping genes for normalizing RNA expression in real-time PCR. BioTechniques **37**(1), 112–119 (2004)
12. Dong, B., et al.: Predicting housekeeping genes based on Fourier analysis. PLoS One **6**(6), e21012 (2011)
13. Eisenberg, E., Levanon, E.Y.: Human housekeeping genes, revisited. Trends Genet. **29**(10), 569–574 (2013)
14. Fayyad, U.M., Piatetsky-Shapiro, G., Smyth, P.: Knowledge discovery and data mining: towards a unifying framework. In: Proceedings of 2nd International Conference on Knowledge Discovery and Data Mining, Portland, OR, pp. 82–88 (1996) https://doi.org/10.1.1.27.363
15. Ghazzali, N.: NbClust: an R package for determining the relevant number of clusters in a data set. J. Stat. Softw. **61**(6), 1–36 (2014)
16. Halkidi, M., Vazirgiannis, M.: Clustering validity assessment: finding the optimal partitioning of a data set. In: Proceedings 2001 IEEE International Conference on Data Mining, pp. 187–194. IEEE (2001)
17. Hall, M., Frank, E., Holmes, G., Pfahringer, B., Reutemann, P., Witten, I.H.: The WEKA data mining software. ACM SIGKDD Explorations **11**(1), 10–18 (2009). https://doi.org/10.1145/1656274.1656278
18. Han, J., Kamber, M., Pei, J.: Data Mining: Concepts and Tecniques, 3rd edn. Morgan Kaufmann/Elsevier, Walthan (2011)
19. de Jonge, H.J.M., et al.: Evidence based selection of housekeeping genes. PLoS One **2**(9), 1–5 (2007). https://doi.org/10.1371/journal.pone.0000898
20. Kovács, F., Legány, C., Babos, A.: Cluster validity measurement techniques. In: Proceedings of the 6th International Symposium of Hungarian Researchers on Computational Intelligence, pp. 1–11 (2005)
21. Kozera, B., Rapacz, M.: Reference genes in real-time PCR. J. Appl. Genet. **54**(4), 391–406 (2013)
22. Lercher, M.J., Urrutia, A.O., Hurst, L.D.: Clustering of housekeeping genes provides a unified model of gene order in the human genome. Nat. Genet. **31**(2), 180–183 (2002). https://doi.org/10.1038/ng887
23. Libbrecht, M.W., Noble, W.S.: Machine learning applications in genetics and genomics. Nat. Rev. Genet. **16**(6), 321–332 (2015). https://doi.org/10.1038/nrg3920
24. Lin, Y., et al.: Evaluating stably expressed genes in single cells. bioRxiv p. 229815 (2018)
25. Liu, P., Si, Y.: Cluster analysis of RNA-sequencing data. In: Datta, S., Nettleton, D. (eds.) Statistical Analysis of Next Generation Sequencing Data. FPSS, pp. 191–217. Springer, Cham (2014). https://doi.org/10.1007/978-3-319-07212-8_10
26. Maimon, O., Rokach, L.: Introduction to knowledge discovery and data mining. In: Maimon, O., Rokach, L. (eds.) Data Mining and Knowledge Discovery Handbook, pp. 1–15. Springer, Boston (2009). https://doi.org/10.1007/978-0-387-09823-4_1
27. Oyelade, J., et al.: Clustering algorithms: their application to gene expression data. Bioinform. Biol. Insights **10**, BBI-S38316 (2016)
28. Pinto, A.C., et al.: Differential transcriptional profile of Corynebacterium pseudotuberculosis in response to abiotic stresses. BMC Genomics **15**(1), 14 (2014)

29. R Core Team: R: A Language and Environment for Statistical Computing. R Foundation for Statistical Computing, Vienna, Austria (2018). https://www.R-project.org/
30. Rao, J., Liu, W., Xie, H.: A new method to identify housekeeping genes and tissue special genes. In: International Conference on Biomedical and Biological Engineering. Atlantis Press (2016)
31. Rendón, E., Abundez, I., Arizmendi, A., Quiroz, E.M.: Internal versus external cluster validation indexes. Int. J. Comput. Commun. **5**(1), 27–34 (2011)
32. Rocha, D.J.P., Santos, C.S., Pacheco, L.G.C.: Bacterial reference genes for gene expression studies by RT-qPCR: survey and analysis. Antonie Van Leeuwenhoek **108**(3), 685–693 (2015). https://doi.org/10.1007/s10482-015-0524-1
33. Ross, I., Gentleman, R.: R: a language for data analysis and graphics. J. Comput. Graph. Stat. **5**(3), 299–314 (1996)
34. Si, Y., Liu, P., Li, P., Brutnell, T.P.: Model-based clustering for RNA-seq data. Bioinformatics **30**(2), 197–205 (2014). https://doi.org/10.1093/bioinformatics/btt632
35. Silva, A., et al.: Complete genome sequence of corynebacterium pseudotuberculosis I19, a strain isolated from a cow in israel with bovine mastitis. J. Bacteriol. **193**(1), 323–324 (2011)
36. Soares, S.C., et al.: Genome sequence of Corynebacterium pseudotuberculosis biovar equi strain 258 and prediction of antigenic targets to improve biotechnological vaccine production. J. Biotechnol. **167**(2), 135–141 (2013). https://doi.org/10.1016/j.jbiotec.2012.11.003
37. Treangen, T.J., Salzberg, S.L.: Repetitive DNA and next-generation sequencing: computational challenges and solutions. Nat. Rev. Genet. **13**(1), 36–46 (2013). https://doi.org/10.1038/nrg3117.Repetitive
38. Vandesompele, J., et al.: Accurate normalization of real-time quantitative RT-PCR data by geometric averaging of multiple internal control genes. Genome Biol. **3**(711), 31–34 (2002). https://doi.org/10.1186/gb-2002-3-7-research0034
39. Vieira, A., et al.: Comparative validation of conventional and RNA-Seq data-derived reference genes for QPCR expression studies of colletotrichum Kahawae. PLoS One **11**(3), e0150651 (2016)
40. Zhao, Y., Wu, J., Yang, J., Sun, S., Xiao, J., Yu, J.: PGAP: pan-genomes analysis pipeline. Bioinformatics **28**(3), 416–418 (2012). https://doi.org/10.1093/bioinformatics/btr655

Predicting Cancer Patients' Survival Using Random Forests

Camila Takemoto Bertolini, Saul de Castro Leite,
and Fernanda Nascimento Almeida[✉]

Federal University of ABC - UFABC, Campus São Bernardo do Campo/SP,
São Bernardo do Campo, Brazil
fernanda.almeida@ufabc.edu.br

Abstract. The increasing amount of data available on the web, coupled
with the demand for useful information, has sparked increasing interest
in gaining knowledge in large information systems, especially biomedical
ones. Health institutions operate in an environment that has been generating thousands of health records about patients. Such databases can be
the source of a wealth of information. For instance, these databases can
be used to study factors that contribute to the incidence of a pathology
and thereby determine patient profiles at the earliest stage of the disease.
Such information can be extracted with the help of Machine Learning
methods, which are capable of dealing with large amounts of data in
order to make predictions. These methods offer an opportunity to trans-
late new data into palpable information and, thus, allows earlier diagnosis
and precise treatment options. In order to understand the potential of
these methods, we use a database that contain records of cancer patients,
which is made publicly available by the Oncocentro Foundation of São
Paulo. This database contains historical clinical information from cancer
patients of the past 20 years. In this paper we present an initial investigation towards the goal of improving prognosis and therefore increasing
the chances of survival among cancer patients. The Random Forest Classification Model was employed in our analysis; this model shows to be
a suitable predicting tool for ours purpose. Thus, we intend to present
means that allows the design of predictive, preventive and personalized
treatments, as well as assisting in the decision making process of the
disease.

Keywords: Cancer database · Machine learning methods · Support
decision-making · Data analysis · Prediction method · Random Forests

1 Introduction

In recent years, research on biological data has increasingly required techniques
that enable data to be collected quickly and with low cost. New record sources
that generate big volume of data are becoming increasingly common and are
getting merged with Hospital Information Systems that have more traditional

© Springer Nature Switzerland AG 2020
L. Kowada and D. de Oliveira (Eds.): BSB 2019, LNBI 11347, pp. 96–106, 2020.
https://doi.org/10.1007/978-3-030-46417-2_9

data, such as electronic health records. However, integrated analysis methods and tools to extract knowledge about this data are not being developed at the same rate as the records are produced. Today, large collections of data with low or no structure can be easily collected from a wide variety of sources, but knowledge about this data has been left behind [11]. Relevant and updated information are crucial to improving decision-making in Healthcare Systems.

The existence of tools capable of storing and making available clinical data in a safe and complete way enables more complete medical record that, in great volume, helps in the development of new diagnostic and treatment techniques. We have come across many studies reporting the potential of Machine Learning (ML) for big data analysis, especially in the medical field [4, 7, 17, 18]. In healthcare, ML applications may offer better indications of the risks and implications of the correlation between diagnosis and therapies; data that may later be confirmed by randomized controlled trials in a sample of patients [5, 17]. ML has been widely applied in the diagnosis and prognosis of certain diseases, mainly cancer patients [9]. Predictive modeling of health data can accelerate research and facilitate healthcare area quality improvement initiatives.

New technologies and computational methods can be the answer to solving problems of the modern healthcare systems, such as the need for patient-physician communication, follow-up appointments, and the availability of specialists. Although such technologies will never completely replace doctors, but they can transform the healthcare industry.

The paper presented by Bhardwaj et al. [1] discusses how a few years ago physicians treated patients based on symptoms; now, these professionals are beginning to diagnose and treat patients with a concept known as evidence-based medicine. This involves reviewing large volume of data from clinical trials and other large-scale treatment pathways and the process of making decisions is based on the best available information. The following example, which was also discussed in [1], gives a real illustration about a patient-physician case and how new technologies can be usefully: if a patient comes in with a particular case of the flu, a physician in the past would rely on what he or she knew about the flu in general or what other doctors in the area knew. With big data technologies, a physician can look at nationwide trends on what course of treatment would work best for the patient to prescribe the best medications. The aggregation of individual data sets that would otherwise prove meaningless provides doctors with the information needed to make better, more holistic medical decisions.

In recent years the accuracy of cancer prediction using ML has increased. There are several studies in the literature that seek the use of numerous ML techniques to detect cancer in its early stage. The choice of method in ML is not a simple choice, as each method can present varied results according to the type of preprocessing applied. The literature survey presented in [9] shows, among the most relevant publications, which ML method was used for the following cases: cancer susceptibility prediction, cancer recurrence within five years after its elimination and the prediction of cancer patient survival. ML can prove to be the solution for both reducing the rising cost of healthcare and helping establish a better patient-physician relationship.

The Hospital Registry of São Paulo Cancer (RHC) is a state database maintained and updated quarterly by the São Paulo Oncocenter Foundation (State Department of Health - São Paulo State Government). There are analytical and non-analytical cases of cancer seen in hospitals in the state of São Paulo since January 2000. After a period of almost 20 years of data acquisition, the São Paulo Oncocenter Foundation raised a wide web database of cancer cases (816,107 as of December 2018) ranging all sort of demographic characteristics and cancer types [14]. The database contains a wide variety of information regarding demographic characteristics of patients as well as several cancer specific classifications that embed a rich medical knowledge in it. This means that, despite the fact that at first glance applying ML algorithm to a large data set might result in a shallow analysis, it is not true, since the data itself carries deep medical information [3]. There are many different prediction purposes that one can have when analyzing this rich database. However, for the sake of simplicity and test of practicability, the patient's survival was chosen to be predicted as a proof of concept.

2   Exploring the Database

An initial exploration of the data set is useful in order to acquire some insights and guidance about the preprocessing that needs to be performed. The patient's age when the cancer is first diagnosed has a slightly inverse correlation (-0.1438) to the chances of survival. The chart on the upper-left corner of Fig. 1 shows that patients diagnosed before the age of approximately 65 years have a higher chance of survival when compared to those diagnosed after this age. In other words, up to the age of 65, the chances of survival for a patient are greater than the chance of death. For instance, there are 6,883 cases of patients diagnosed at the age of 40 years, 4,672 alive and 2,211 deceased. In general, female patients tend to be diagnosed with cancer earlier than their male counterparts. The chart on the upper-right corner of Fig. 1 shows the age of diagnosis per gender for the years of 2014 until 2018. Also, there are types of cancer with higher chances of occurrence depending on the age of the patient. Figure 1 (bottom) makes this situation explicit for both genders.

In addition, the data set provides four possible categories for the patients' last know status: "alive without cancer," "alive with cancer," "deceased by cancer," and "deceased by other reason." All 4 categories together with their respective occurrence percentages may be seen in the chart of Fig. 2. If "alive without cancer" and "alive with cancer" are grouped under a single category "alive" (same thing done to "deceased by cancer" and "deceased by other reason"), there is a small lack of balance between the number of patients alive and deceased. In numbers, 57.65% of the patients are alive while 42.35% are deceased.

3   Machine Learning Application

Generally, the ML framework steps are as follows: preprocessing, training, and testing. In the preprocessing step is intended to highlight important features existing in the data set, and (at the same time) to do a cleanup, removing entries that not so important and that may interfere with the learning algoritm.

It is important that the procedural steps defined for this purpose could be applied to totally new data that can be presented in the future. The training step consists of finding the best algorithms, parameters, and data features that

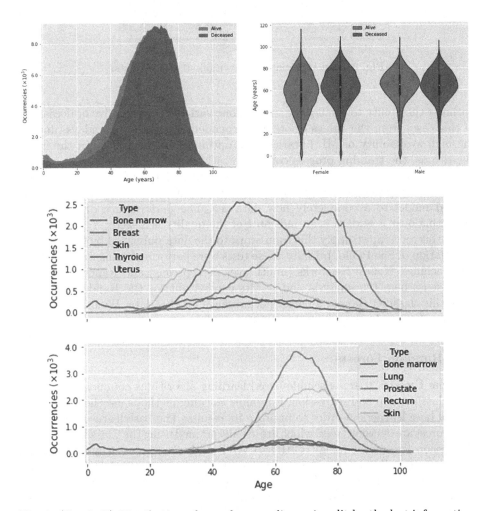

Fig. 1. (Top-Left) Distribution of age of cancer diagnosis split by the last information about the patient (alive or deceased). (Top-Right) Distribution of age of cancer diagnosis split by gender and the last information about the patient (alive or deceased) for cases diagnosed between 2014 and 2018. (Bottom) Distribution of most frequent types of cancer by age and gender.

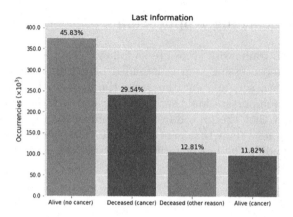

Fig. 2. Last information known about the patients.

optimize the ability to predict the value of one variable (target: alive or deceased, in this case) from the values of other independent variables. Nowadays, due to the high availability of ML libraries with optimal implementation of the main algorithms, the training process consists mostly of trial and error using several combinations of parameters and data subsets, having the insights from exploration and preprocessing as guidance. The testing step is the application of the trained model to a completely new data set. The performance of the model predicting over new data is an indicator of the quality achieved.

In this section, we apply this framework to the data published by Oncocentro Foundation of São Paulo. Initially, these tests were performed using Support Vector Machines, Neural Networks (multi-layer perceptrons) and Random Forests. We discuss here the analysis and results carried out for the Random Forests, since this classifier showed a better performance for this data. We briefly review Random Forests in the next subsection.

3.1   Random Forests

Random Forests [2] are ensemble-based learning algorithms commonly used for data classification and regression. Ensemble learning methods use the decision of several base learners in order to improve its results. The base learners of a Random Forest are decision trees, which are usually trained using the Classification and Regression Tree methodology (CART). CART methods generate a binary decision tree by recursively dividing the tree nodes into child nodes containing an increasingly more homogeneous split of the data [6]. However, the decision trees used in Random Forests are grown non-deterministically in a two-stage randomization procedure [6]. The randomization procedures are usually implemented by training each tree with a different sample of the original data and, in addition, instead of splitting a tree node using all features, the Random Forest selects on each node of each decision tree a random subset of features. Only these features are used as candidates to find the best division for the node [6]. Thus, the Random Forest classification method consists of a classifier that contains a collection of tree-structured classifiers, where each tree in grown in a randomized way. Each decision tree is then used to contribute with an unit vote in the decision process [8].

The main advantages of Random forests are that they are capable of handing non-linear data, are simpler to tune, and have an intrinsic feature selection step [15]. This feature selection step can be important in determining the most informative features in a decision problem.

3.2 Preprocessing

The data set contains dozens of columns. For this study, we have chosen the following attributes in the analysis, which are show divided by their types:

(a) Categorical: educational level, gender, the method used for a diagnostic, facility where the patient was diagnosed, type of cancer, level of development of the primary tumor, level of spreading of the tumor in the surrounding tissues, and level of spreading of the tumor in other areas;
(b) Numerical: age;
(c) Binary: if the patient went through surgery if the patient went through radiotherapy if the patient went through chemotherapy if the patient went through hormone therapy;
(d) Dates: date of the first visit to a physician, date of the diagnostic, the date of the treatment's beginning.

The following modifications were performed in the data extracted from the original set in order to improve the analysis. First, instead of using the dates directly, they were replaced by the difference in days between each other. This resulted in simple numbers that were used by the algorithms. The categorical variables were replaced by groups of dummy variables. For instance, "method used for diagnostic" is a variable that accepts only 4 different values: "clinical exam", "microscopic resources", "non microscopic resources", and "not available". This column was replaced by 4 columns accepting values 1 or 0 (true or false). Therefore, if "method used for diagnostic" has the value "clinical exam", this column will have the value 1, while the other three columns will have the value 0.

In addition, since there is a lack of balance between the number of cases where the patient is alive and deceased, we randomly removed cases where the patient is alive until the number of cases in each category is equal. This was performed in order to improve classification results.

The procedure adopted for preprocessing can be summarized in following steps: (i) filter only chosen columns; (ii) replace dates by the difference in days between each other; (iii) replace categorical variables by groups of dummy variables; and (iv) for the training data: randomly remove cases where the patient is alive until the number of cases in each category is equal. All these steps are carried using the Python Pandas Library [10].

3.3   Training and Testing

As it is common in ML, in order to test our classification algorithms, the data set was randomly split into training and test sets. The training set contained 75% of the preprocessed data. The remaining 25% was used for testing.

We used the Python Scikit Learn Library for the implementation of the Random Forest Classifier, as well as another tool that are necessary to make the best use of the trained model, since it provides industry quality implementation of this classifier. For instance, Scikit Learn provides a Grid Search and Cross-Validation functionality [12]. Grid Search is a procedure for finding the model's parameters that lead to the best results. However, it is known that training a model seeking to optimize accuracy can result in over-fitting [16]. Over-fitting can happen when a model is not very good in generalization because it is strongly held to the training data specific characteristics. When this happens, the model might present great results when used to predict known data, but very poor ones when it comes to new data.

The most widely used technique to avoid over-fitting is Cross-Validation. There are many variations, but the idea is that the training data is not used only once to train the model. For instance, in the K-Fold procedure, data is split in K subsets $\{D_1, D_2, \ldots, D_k\}$; the model is trained with $\{D_1, D_2, \ldots, D_{k-1}\}$ and tested against D_k. After the first training, the model is trained with $\{D_1, D_2, \ldots, D_{k-2}, D_k\}$ and tested against the remaining D_{k-1} set, so on. In the end, instead of having one single expected accuracy, there are K expected ones, that can be averaged in order to have a better understanding of the model's ability to generalize its procedure.

In order to perform both the Grid Search and the Cross Validation procedures, we used the `GridSearchCV` class of the Scikit Learn Library. The metrics used to track the models quality were the following [13]:

(a) Accuracy: the ratio between correct classifications and the total of cases. It is given as a percentage;
(b) $F1$: a metric that combines many ratios indicating relations between true positives, true negatives, false positives, and false negatives. It is the average of $F1_+$ and $F1_-$. It ranges between 0 and 1 (or 0 and 100 %) and as bigger its value, the better is the model's quality;
(c) $F1_+$: the $F1$ score regarding positive cases, which means the capacity of the model in predicting a positive case (the patient is alive);
(d) $F1_-$: the $F1$ score regarding negative cases, which means the capacity of the model in predicting a negative case (the patient is deceased).

The following parameters were considered for the Random Forest Classifier and were optimized by the Grid Search procedure described above: `n_estimators`, which gives the numbers of trees in the forest, was tested values: [5, 10, 15, 20, 25, 30]; `min_samples_split`, which gives the minimum number of samples required to create a split in a tree, was tested with values: [2, 20, 200]; `max_features`, the maximum number of columns (or features) to be considered when creating a split in a tree, was tested with values: ["auto", "sqrt", "log2"]; and warm_start, which tells the model to start with all trees already in place (False) or to add trees during the training (True), was tested with values: [True, False]. The best values found for the parameters above were: n_estimator with value of 30, min_samples_split set to 20, max_features set to sqrt and warm_start set to True. The quality the model trained with 75% of the data using the parameters above and tested against the remaining 25% is given in Table 1 under the column **OPT1**.

Since the best value for the parameter n_estimators was the one with the highest value, another Grid Search (assisted by Cross Validation) was performed using the following parameters: n_estimators varying in the set [30, 50, 100]; min_samples_split varying in the set [20, 30, 40]; and max_depth varying in the set [None, 5, 10, 15, 20, 25, 30]. This last parameter limits the maximum number of levels each tree might have. When None is used, the number of levels is limited by the minimum number of samples required to create a split in a tree (min_samples_split). The best values found for the parameters above were: n_estimators as 100; min_samples_split set to 20; and max_depth set to None, which is the model's *default* value, so nothing changes in relation to the first optimization. The quality of a model trained with 75% of the data using the parameters above and tested against the remaining 25% is displayed by Table 1 under the column **OPT2**.

Table 1. Model Quality Comparisons using the test set. Default represents the results for the Random Forest algorithm running with default values and **OPT1** and **OPT2** show the algorithm running the optimal values found by the grid search method.

Metrics	Default	OPT1	OPT2
Accuracy	72.60 %	75.11 %	75.36 %
$F1$	72.59 %	75.00 %	75.25 %
$F1_+$	73.12 %	76.64 %	76.90 %
$F1_-$	72.05 %	73.36 %	73.59 %

Since the best value for n_estimators was again the largest one tested, this seems to indicate that the effectiveness of the Random Forest Classifier increases with the number of trees used. However, this comes also with an increase of computational cost. In order to show how this computational cost increases with the number of trees, we performed a test using the best parameters found above, but using different values of n_estimators ranging from 5 to 30. The result of this analysis is shown in Fig. 3.

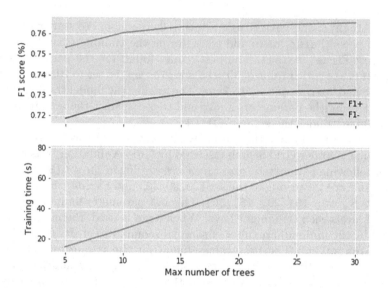

Fig. 3. Relation between the model's quality ($F1$ score) and the maximum number of trees, as well as the time necessary to train the model.

4   Conclusion

In this paper we perform an initial investigation using the publicly available data from Oncocentro Foundation of São Paulo using the Random forest classifier implemented by the Python Scikit Learn Library. It was shown that one can rapidly obtain a classifier that can be used to predict whether a person is going to survive cancer with 75.25% of accuracy.

As a prototype, this analysis is a good starting point for making clear what are the open source tools available, what are the possible challenges when it comes to the necessary computing power, and what are the steps required in order to train an initial ML model. In this analysis, not all steps were executed, some of them were omitted since it was possible to initially move forward without them. As examples of steps that worth to be mentioned:

– the initial exploration of the data can be performed with more detailed statistical analysis. This could result in insights that could guide the model's training in order to avoid traps and improve quality;
– the use of a group of dummy variables to replace categorical ones flood the data set with dozens of columns, it not only increases the required computing power but also the complexity of the data. This is a perfect situation where a feature selection or dimensionality reduction methods, such as the Principal Component Analysis, could be used.

The ultimate goal of such these ML tools is to assist health care professionals to identify patients that can have their survival chances increased. In order to pursue such a goal, a much deeper analysis is required, taking into account all the medical information available in the data set, but also making use of the chronological events. For instance, it is known that the database is updated quarterly, even not being possible to uniquely identify patients, it is possible to track the evolution of determined groups of individuals across the data set updates.

Acknowledgments. This article was supported in part by CECS/UFABC and FAPESP (Process Number 2019 / 21613-7)

References

1. Bhardwaj, R., Nambiar, A.R., Dutta, D.: A study of machine learning in healthcare. In: 2017 IEEE 41st Annual Computer Software and Applications Conference (COMPSAC), vol. 2, pp. 236–241 (2017). https://doi.org/10.1109/COMPSAC.2017.164
2. Breiman, L.: Random forests. Mach. Learn. **45**(1), 5–32 (2001). https://doi.org/10.1023/A:1010933404324
3. Cancer.gov: National cancer institute: Cancer staging (2015). https://www.cancer.gov/about-cancer/diagnosis-staging/staging. Accessed 16 Apr 2019
4. Chen, M., Hao, Y., Hwang, K., Wang, L., Wang, L.: Disease prediction by machine learning over big data from healthcare communities. IEEE Access **5**, 8869–8879 (2017)
5. Chen, P.H.C., Liu, Y., Peng, L.: How to develop machine learning models for healthcare. Nat. Mater. **18**(5), 410–414 (2019). https://doi.org/10.1038/s41563-019-0345-0
6. Chen, X., Ishwaran, H.: Random forests for genomic data analysis. Genomics **99**(6), 323–329 (2012)
7. Dua, S., Acharya, U.R., Dua, P. (eds.): Machine Learning in Healthcare Informatics. ISRL, vol. 56. Springer, Heidelberg (2014). https://doi.org/10.1007/978-3-642-40017-9
8. Eesha Goel, E.A.: Random forest: a review. Int. J. Adv. Res. Comput. Sci. Softw. Eng. **7**, 251 (2017). https://doi.org/10.23956/ijarcsse/V7I1/01113
9. Kourou, K., Exarchos, T.P., Exarchos, K.P., Karamouzis, M.V., Fotiadis, D.I.: Machine learning applications in cancer prognosis and prediction. Computat. Struct. Biotechnol. J. **13**, 8–17 (2015). https://doi.org/10.1016/j.csbj.2014.11.005
10. McKinney, W.: Data structures for statistical computing in python. In: van der Walt, S., Millman, J. (eds.) Proceedings of the 9th Python in Science Conference. pp. 51–56 (2010)
11. Ohno-Machado, L.: Data science and informatics: when it comes to biomedical data is there a real distinction? J. Am. Med. Inform. Assoc. **20**(6), 1009–1009 (2013). https://doi.org/10.1136/amiajnl-2013-002368
12. Pedregosa, F., et al.: Scikit-learn: machine learning in Python. J. Mach. Learn. Res. **12**, 2825–2830 (2011)
13. Powers, D.: Evaluation: from precision, recall and f-measure to roc, informedness, markedness & correlation. J. Mach. Learn. Technol. **2**, 2229–3981 (2011). https://doi.org/10.9735/2229-3981
14. RHC: Fundação oncocentro de são paulo. http://www.fosp.saude.sp.gov.br/publicacoes/downloadarquivos. Accessed 15 Apr 2019

15. Sarica, A., Cerasa, A., Quattrone, A.: Random forest algorithm for the classification of neuroimaging data in alzheimer's disease: a systematic review. Front. Aging Neurosci. **9**, 329 (2017). https://doi.org/10.3389/fnagi.2017.00329
16. Scikit-learn.org: Tuning the hyper-parameters of an estimator. https://scikit-learn.org/stable/modules/grid_search.html. Accessed 18 Apr 2019
17. Seyhan, A.A., Carini, C.: Are innovation and new technologies in precision medicine paving a new era in patients centric care? J. Transl. Med. **17**(1), 114 (2019). https://doi.org/10.1186/s12967-019-1864-9
18. Wiens, J., Shenoy, E.S.: Machine learning for healthcare: on the verge of a major shift in healthcare epidemiology. Clin. Infect. Dis. **66**(1), 149–153 (2017)

Extended Abstracts

Searching *in Silico* Novel Targets for Specific Coffee Rust Disease Control

Jonathan D. Lima[1], Bernard Maigret[5], Diana Fernandez[3,4],
Jennifer Decloquement[2], Danilo Pinho[2], Erika V.S. Albuquerque[3],
Marcelo O. Rodrigues[1], and Natalia F. Martins[6(✉)]

[1] Inorganic and Materials Laboratory, University of Brasília - UnB, Brasília, Brazil
[2] Department of Plant Pathology (IB/UnB), University of Brasília - UnB,
Brasília, Brazil
[3] Embrapa Genetic Resources and Biotechnology, Brasília, Brazil
[4] IRD, Cirad, Univ Montpellier, IPME, Montpellier, France
[5] CNRS, LORIA, UMR 7503, Lorraine University, Nancy, France
[6] Embrapa Agroindustria Tropical, Fortaleza, CE, Brazil
natalia.martins@embrapa.br

Abstract. Coffee industry is threatened by production losses due to the rust disease since 1850. The coffee leaf rust (CLR) disease has an important social and economical impacts and its control is still a challenge. The CLR pathogen is the fungus *Hemileia vastatrix* Berkeley & Broome (*Basidiomycota*, order *Pucciniales*) that is currently controlled by using non-specific anti-fungal chemicals spraying. The advances in molecular biology and bioinformatics may allow the identification of new targets and environmentally safe strategies for controlling CLR. Several genomic and transcriptomic data are available for *H. vastatrix* that allow searching for new proteins to achieve a better disease control. We used the dataset of 34,242 sequences from the fungal genome and transcriptome, with a filtering strategy for protein annotation, structure and cell sublocalization to select three essential proteins related to steroid synthesis, cell membrane, and cell wall metabolism. This short paper reports the ongoing study to allow the development of new molecules that might be validated and contribute to new products that are specific and ecologically friendly.

Keywords: *Hemileia vastatrix* · Sequence analysis · Target search

1 Introduction

Coffee is the world's second most valuable traded commodity. According to the International Coffee Organization (ICO) [1], coffee production in the world during 2018/19 increased by an estimated 1.9% to 169 million bags, led by an 18.5% increase in Brazilian production, making Brazil the world's largest producer.

Supported by Bayer Grants4Target/Embrapa.

Plant diseases are responsible for significant losses in crop production and their control is extremely important for the coffee production. One of the main coffee (*Coffea arabica* L.) diseases is coffee leaf rust (CLR) which causes losses of over US$ 1 billion per year and interferes with the production of arabica coffee [2]. The disease symptoms are leaf discoloration and presence of orange pustules on the lower leaf surface [3]. Severe symptoms can cause leaf fall and repeated disease cycles can lead to tree death. The CLR pathogen is the fungus *Hemileia vastatrix* Berkeley & Broome (*Basidiomycota*, order *Pucciniales*) first described in 1869 in Ceylan where it destroyed all coffee plantations [4]. CLR has caused increasing losses in coffee production in Africa and Asia and appeared in Brazil in 1970, probably transported through the wind high currents [3].

Search for natural resistance to CLR and breeding for resistant coffee varieties have been the focus of research for decades [3,5]. However, genetic control of CLR has been hampered by great pathological diversity and rapid genetic evolution of the fungus overcoming plant resistance genes deployed so far [2]. CLR control may be efficiently achieved through chemicals use, including copper-based fungicides and others with different chemical forms [6]. However, the environmental toxicity and non-selectivity of these chemical compounds must be taken into consideration. The search for molecules and the design of drugs that can act on pathogen-specific molecular targets without promoting toxicity to the host and the environment are some of the goals for controlling plant pathogens nowadays.

The advances in molecular biology and bioinformatics may allow the identification of new targets and strategies for controlling plant diseases [7]. With the growth of information coming from genomics and transcriptomics, the understanding of biological systems has increased as well as biotechnological applications [7,8]. Several genomic and transcriptomic resources are available for *H. vastatrix* [9–12], that allow searching for essential target proteins expressed during germination of urediniospores and subsequent plant infection. Based on these important data it should possible to propose and validate new approaches to develop novel specific and safer drugs to control CLR.

2   Methods

To rank and select *H. vastatrix* proteins, we screened 6,763 proteins from the *H. vastatrix* transcriptome when infecting leaves [9], 14,445 from the genomic data set [11] and 13,034 putative proteins with transcriptomic support from [12]. Figure 1 shows the strategy for the selection of candidate proteins.

Thus, the identification of potential targets was based on the ensemble of sequences coming from the fungi genome, assembled transcriptome sequences. The BLASTx search was performed for the total amount of sequences to characterize their functions. Additionally, another BLASTx search was performed against three specific datasets: the predicted secretome, PHI-base version 5.0 and a group of known protein targets selected by fungal enzyme inhibitors [13].

As our primary objective was to identify genes that could be used as targets for the development of new forms to control the disease, our systematic criteria were: (1) select protein annotation, phenotype description, and transcriptomic evidence; (2) filter repetitive sequences, proteins with low structural similarity within the Protein Databank (PDB), cell localization prediction in nuclei, and candidates with low accessibility to chemical compounds; (3) protein molecular size; (4) the possible target selection also considered the absence of orthologues in another organism such as insects, plants, humans and evolutionary conservation among other fungi. Therefore, our selection process consisted of several elimination rounds. We kept as representative three proteins for further analysis.

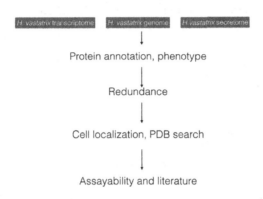

Fig. 1. Strategy of protein target selection.

Protein Modeling. Homology modeling is a well known method to reach reasonable theoretical 3D models. To build 3D models of the three selected targets we used the SwissProt modeling server [14]. Once the target protein was identified as the most suitable template for homology modeling, we used MUSCLE 3.8.31 for multiple sequence alignment with default parameters to check for sequence similarity and to verify the conservation of structural signatures. The crude model for each selected target was obtained and equilibrated by simulated annealing.

Ligand Docking and Virtual Screening. To find new leads for the development of new fungicides, we performed a molecular docking protocol with a previously selected chemical library [7]. The parameters used were direct docking into the active site of each enzyme using protonation and hydrogen for all molecules in the DOCKTHOR server [15].

Experimental Validation. To experimentally validate the chemical compounds designed based on the structure-function relationship of *H. vastatrix* protein targets, solubility tests and rust bioassays were performed with 10 synthesized chemical compounds (Life Chemicals Worldwide). Urediniospores of *H. vastatrix* race II were produced on *C. arabica* Catuai Amarelo grown in a greenhouse during the wet season (November to April) in Brasilia. To evaluate the inhibition of *H. vastatrix* urediniospore germination, tests were performed on coffee leaf disks inoculated with ca. 100 urediniospores in a 30 μL aqueous solution containing the chemical compound at its highest solubility concentration. Disks inoculated with urediniospores in water were used as germination control. Petri dishes filled with water agar (2%) were used to condition the leaf disks (abaxial side up) and kept at 26 °C in darkness. After 24 h, leaf disks were briefly immersed in a 0.1% solution of Fluorescent Brightener (Sigma-Aldrich) to stain the fungal structures, and spore germination was evaluated by observation under a Zeiss AxioPhot fluorescence microscope equipped with a DAPI filter. The percentage of spore germination was calculated based on the number of germinated spores to the total number of spores on each leaf disk. Each treatment had triplicates and the final germination result was obtained by average triplicate data.

3   Results and Discussion

The growth of genomic data allows data mining and the discovery of specific targets in organisms. This approach has been used for several plants and human diseases [16]. Now, the prediction of protein structure-function relationship reveals, from its structure, the binding of drugs at a specific location, just as a key and lock search. The discovery of new antifungals is still challenging because as eukaryotes, fungi contain relatively few exclusive targets for developing structure-based compounds.

According to the selection strategy, described in Fig. 1, we selected the proteins with a lethal phenotype, considering that once inhibited the target could change the life cycle of the pathogen, causing a lower level of infection or even cell death (Table 1). Therefore we selected three candidates, the mevalonate kinase, a Δ9-acyl-CoA desaturase, and the (1–3)-β-D-glucan synthase. Nevertheless, not all predicted proteins from *H. vastatrix* were submitted to the strategy of selection of candidates to virtual drug screening, due to lack of 3D structure.

The mevalonate pathway plays a key role in multiple cellular processes by synthesizing sterol isoprenoids, such as cholesterol, and non-sterol isoprenoids. It is also known as the isoprenoid pathway or HMG-CoA reductase pathway is an essential metabolic pathway present in eukaryotes, archaea, and some bacteria. Among the enzymes, in the pathway, the mevalonate kinase plays a key role in isoprenoid and sterol synthesis. This enzyme is also known as Ergosterol biosynthesis protein 12 (ERG12). As Ergosterol is a unique component of fungal cells, it has been reported as a feasible target for fungicides. Recent studies in *Aspergillus oryzae* transcriptome revealed that the Ergosterol pathway is sensitive to terbinafine (target site is ERG1) and tebuconazole (target site is ERG11) [17]. Furthermore, the comparison with proteins orthologues might reveal important conservation in structure-sequence function.

Table 1. Summarized results for target selection in *H. vastatrix* databases [9–12].

Seq name	Putative protein function	Phenotype	Cell localization	Protein size	PDB search
Supertig03605	Energy production and conversion	Lethal	Cytoplasm	228 aa	50% (6C45)
Supertig01121	Posttranslational modification	Lethal	Cytoplasm	290 aa	80% (5GJQ)
Supertig00354	Lipid transport and metabolism	Lethal	Cytoplasm	305 aa	No match
Supertig00220	Polyadenylate-binding protein	Lethal	Nuclei	218 aa	56% (6R5K)
Supertig01025	Lipid transport and metabolism	Lethal	Secreted	237 aa	36% (3KSO)
Supertig01528	Cell wall/membrane/ envelope biogenesis	Lethal	Cytoplasm	324 aa	No hits
Supertig00499	Posttranslational modification	Lethal	Cytoplasm	490 aa	58% (6NYY)
Supertig00041	RNA processing and modification	Lethal	Cytoplasm	307 aa	44% (2MJN)
Supertig04243	Translation, ribosomal structure and biogenesis	Lethal	Cytoplasm	788 aa	48% (4YE9)
Supertig00182	Cytoskeleton	Lethal	Cytoplasm	431 aa	72% (1FFX)
Supertig00335	RNA-binding protein	Lethal	Chloroplast	101 aa	46% (2MPU)

In animal and fungal cells, the fatty acid plays an important role as structural components of membrane phospholipids. They are present in all groups of organisms, i.e., bacteria, fungi, plants and animals, and play a key role in the maintenance of the proper structure and functioning of biological membranes. Although all membranes have similar composition, the Δ9-acyl-ACP desaturases of higher plants are structurally different from the Δ9-acyl-CoA desaturases of yeast and mammals [18]. The inhibition of this particular enzyme might lead to loss of membrane structure and lethality of the pathogen. For this protein target, the modeling revealed some structural features of the binding site but a great amount of the structure is missing (Fig. 2). Further mining at the genome might complete the sequence and therefore improve the 3D modeling.

Currently, antifungal agents such as triazoles, amphotericin B and echinocandins block the synthesis of the plasma membrane disrupting the fungal plasma membrane; blocking the synthesis of both DNA and RNA, and inhibiting the enzyme responsible for the synthesis of the $(1$–$3)$-β-D-glucan present in the fungal cell wall [19]. Although the phenotype and the annotation for the three candidates were in accordance with the strategy, we were unable to build a suitable model for virtual screening for all targets, once the coverage of the structural template was 20%. The compounds we propose as leads to new fungicides are non-toxic for the environment and other hosts such as insects and humans. Further studies are needed to clarify and improve the activity and to determine the inhibition (Table 2).

Mevalonate kinase
Template 2r3V 47.17% identity

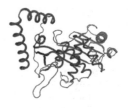

Acyl-CoA desaturase 1
Template 4mymk 29.63% identity

Fig. 2. 3D structure for selected proteins target.

Table 2. Experimental validation of 10 chemical compounds.

Drug	MW	Concentration (mM)	Water dissolution	Germination rate
#1	709	14	1/800	80%
#2	419	24	1/300	80%
#3	514	19	1/500	80%
#4	418	24	1/500	80%
#5	493	20	1/400	80%
#6	451	22	1/200	80%
#7	444	22	1/300	80%
#8	446	22	1/700	80%
#9	434	23	1/200	80%
#10	572	17	1/600	90%
Control	-	-	-	80%

4   Conclusion

This short paper reports the progress in an innovative approach integrating genomics, structural bioinformatics and drug design for proposing new solutions to the coffee rust threat.

References

1. ICO, International Coffee Organization: Coffee Market Report - July 2019, London (2019). http://www.ico.org/documents/cy2018-19/cmr-0719-e.pdf
2. Talhinhas, P., et al.: The coffee leaf rust pathogen Hemileia vastatrix: one and a half centuries around the tropics. Molecular Plant Pathol. **18**, 1039–1051 (2017)

3. Rodrigues Jr., C.J., Bettencourt, A.J., Rijo, L.: Races of the pathogen and resistance to coffee rust. Annu. Rev. Phytopathol. **13**, 49–70 (1975)
4. Berkeley, M.J., Broome, C.E.: Hemileia vastatrix. Gardeners' Chronicle **6**, 1157 (1869)
5. Bettencourt, A.J., Rodrigues Jr., C.J.: Principles, practice of coffee breeding for resistance to rust, other diseases. In: Clarke, R.J., Macrae, R. (eds.) Coffee. Agronomy, vol. 4, pp. 199–234. Elsevier, London (1988)
6. Zambolim, L.: Current status and management of coffee leaf rust in Brazil. Tropical Plant Pathol. **41**(1), 1–8 (2016). https://doi.org/10.1007/s40858-016-0065-9
7. Bresso, E., Togawa, R., Hammond-Kosack, K., Urban, M., Maigret, B., Martins, N.F.: GPCRs from Fusarium graminearum detection, modeling and virtual screening - the search for new routes to control head blight disease. BMC Bioinform. **17**(18), 463 (2016)
8. Verli, H.: Bioinformática: da Biologia à Flexibilidade Molecular. São Paulo: SBBq, 1ª ed (2014)
9. Fernandez, D., et al.: 454-pyrosequencing of Coffea arabica leaves infected by the rust fungus Hemileia vastatrix reveals in planta expressed pathogen-secreted proteins and plant functions triggered in a late compatible plant-rust interaction. Mol. Plant Pathol. **13**, 17–37 (2012)
10. Talhinhas, P., et al.: Overview of the functional virulent genome of the coffee leaf rust pathogen Hemileia vastatrix with an emphasis on early stages. Front. Plant Sci. Plant-Microbe Interact. **5**(88), 1–17 (2014)
11. Cristancho, M.A., et al.: Annotation of a hybrid partial genome of the coffee rust (Hemileia vastatrix) contributes to the gene repertoire catalog of the Pucciniales. Front. Plant Sci. **5**, 594 (2014)
12. Porto, B.N., Caixeta, E.T., Mathioni, S.M., Vidigal, P.M.P., Zambolim, L., Zambolim, E.M.: Genome sequencing and transcript analysis of Hemileia vastatrix reveal expression dynamics of candidate effectors dependent on host compatibility. PLoS ONE **14**(4), e0215598 (2019)
13. Ramakrishnan, J., Rathore, S.S., Raman, T.: Review on fungal enzyme inhibitors - potential drug targets to manage human fungal infections. RSC Adv. **6**(48), 42387–42401 (2016)
14. Waterhouse, A., et al.: SWISS-MODEL: homology modelling of protein structures and complexes. Nucleic Acids Res. **46**(W1), W296–W303 (2018)
15. da Silveira, N.J.F., Pereira, F.S.S., Elias, T.C., Henrique, T.: Web services for molecular docking simulations. In: de Azevedo, W. (eds) Docking Screens for Drug Discovery. Methods in Molecular Biology, vol. 2053 (2019)
16. Sowjanya, K., Girish, C.: Structural genomics in drug discovery: an overview. J. Pharmacol. Pharmacother. **10**, 1–6 (2019)
17. Hu, Z., et al.: Gene transcription profiling of Aspergillus oryzae 3.042 treated with ergosterol biosynthesis inhibitors. Braz. J. Microbiol. **50**, 43–52 (2019)
18. Sakuradani, E., Kobayashi, M., Shimizu, S.: 9-Fatty acid desaturase from arachidonic acid-producing fungus. Euro. J. Biochem. **260**, 208–216 (1999)
19. Perfect, J.R., Tenor, J.L., Miao, Y., Brennan, R.G.: Trehalose pathway as an antifungal target. Virulence **8**(2), 143–149 (2017)

A Domain Framework Approach for Quality Feature Analysis of Genome Assemblies

Guilherme Borba Neumann$^{(\boxtimes)}$, Elvismary Molina de Armas,
Fernanda Araujo Baiao, Ruy Luiz Milidiu, and Sergio Lifschitz

Pontifícia Universidade Católica do Rio de Janeiro (PUC-Rio), Rio de Janeiro, Brazil
{gneumann,earmas,milidiu,sergio}@inf.puc-rio.br,
fbaiao@puc-rio.br

Abstract. The Genome Assembly research area has quickly evolved, adapting to new sequencing technologies and modern computational environments. There exist many assembler software systems that consider multiple approaches; however, at the end of the process, the assembly quality can always be questioned. When an assembly is accomplished, one may generate quality features for its qualification. Nonetheless, these features do not directly explain the assembly quality; instead, they only list quantitative assembly descriptions. This work proposes GAAF (Genome Assembly Analysis Framework), a domain framework for the feature analysis post-genome Assembly process. GAAF works with distinct species, assemblers, and features, and its goal is to enable data interpretation and assembly quality evaluation.

GitHub: https://github.com/neumannguib/GAAF-Framework.

Keywords: Genome Assembly · Feature analysis · Evaluation metrics · Genomics · Quality analysis

1 The Quality Problem of Assemblies

We have been living a considerable revolution in Life Sciences since the advent of Information Technology, and in particular in Genomics. As technology evolves, new data structures come into play, and IT systems need to respond to this evolution. The fields of Genome Sequencing and Genome Assembly face the problem of quality. Current sequencing technology generates a significant number of pieces of DNA that are submitted to assembly algorithms to create an assembly as close as possible to the original molecule. However, the distance between real genomes and assemblies is still an open issue.

High-quality assemblies are expensive and hard to achieve. Therefore, when applications do not strictly require high-quality, scientists leave assemblies at a level where they are called *drafts* — they are not complete, in comparison to

Supported by National Council for the Improvement of Higher Education (CAPES).

L. Kowada and D. de Oliveira (Eds.): BSB 2019, LNBI 11347, pp. 116–122, 2020.
https://doi.org/10.1007/978-3-030-46417-2_11

reference genomes. In GenBank, for example, more than 80% of all bacterium genomes are considered drafts [10]. Analyzing the Genomes OnLine Database (www.gold.jgi.doe.gov), we found 121,000 permanent drafts, in contrast to only 4,000 complete assemblies (on 14th June 2018). But to say whether it is a problem, we need to know the quality of the drafts.

On the one hand, Denton et al. have found extensive errors in the number of genes inferred from draft genome assemblies, mainly due to fragmentation of genes [5]. On the other hand, Land et al. reported that 88% of all GenBank bacterium draft assemblies are good enough according to some thresholds in contiguity and base analysis metrics [10]. For example, genome annotation, during Synteny Analysis (to an error rate below 5%), requires a minimum N50 of 200 kb and 1 Mb when gene density is 290 and 200 genes per Mb, respectively [12].

The quality metrics, extensively provided as the main thresholds for data selection, perform a crucial post-genome assembly step, affecting all the subsequent genomic study applications. For that reason, not only metrics but also features[1] in general, emerged in the last years to qualify assemblies [1,2,4,7,9,13,15,16].

Currently, hundreds of features are available for assembly's description but falling at the problem of interpretation. The biological meaning of a feature is always an analysis up to the biologist. Some of them may be redundant, and others may not represent reality [14]. In light of this context, we introduce a Framework Approach for the process of Genome Assembly Evaluation, providing support for the analysis of quality by end-users.

2   Framework Approach

In this paper, we present a domain framework for the area of Genome Assembly Quality Analysis, named GAAF (Genome Assembly Analysis Framework). The main goal is to allow features comparison, considering distinct sequencing, assembling, and genomic variables. The framework permits, through customized experiment design, the detection of sequencing, and assembly parameters, and also enables the identification of genomic characteristics that affect the quality assembly features. Consequently, our solution contributes to the selection of features which better describe real assembly quality.

Generally, a framework is defined based on four main requirements: code reuse, abstraction, inversion control, and extensibility [11]. According to Landin and Nikalasson, "an object-oriented framework is intended to capture the functionality common to several similar applications" [11]. The requirements of a framework include those typical from all applications and the adaptability for the particular ones. Here, we consider applications related to quality studies for Genome Assemblies.

In addition, the requirements are consequently translated into cold or frozen spots and hot spots. Frozen spots compose the core of the framework, immutable

---

[1] A Feature is a measurable property or characteristic that describes an object or phenomenon under observation [3].

code common to all applications. Contrarily, the hot spots are the changeable modules, the ones that may be absent, and those that may be often added to the application [6].

2.1   Requirements

The proposed GAAF framework is a domain framework for the quality analysis of the Genome Assembly, and must:

(1) handle as many types of assemblers as possible;
(2) work with multiple sequencing technology types;
(3) test distinct genomic characteristics;
(4) test distinct assembling parameters;
(5) generate multiple features;
(6) be able to add assembler, feature, or read generator tools;
(7) be able to work with external raw reads or to generate their reads; and
(8) be able to test hypotheses concerning the whole experiment.

Additionally, non-functional requirements include the independence of the modules, the flexibility to reuse data, and the possibility to partial runs. Any other requirement may be part of the application and work as a hot spot.

2.2   Architecture

We introduce the proposed architecture of GAAF in Fig. 1. The blue color refers to cold spots and the red color to hot spots. We divide the architecture into layers (communication, control, service, and data). The components in each layer are independent of each other and explained below.

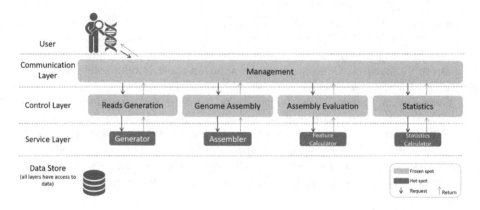

Fig. 1. Framework architecture (Color figure online)

Management: all configuration, inputs, and outputs are managed through the main core module called Management. It works as the main line of communication between all other frozen spots, and between User and GAAF. The Management core and the other frozen spots persist metadata for each run, granting the possibility to posterior partial executions. When errors occur, new tools are added, or the execution is paused, the metadata is used for partial runs.

Reads Generation: as the name mentions, Reads Generation Module is responsible for generating artificial reads according to the chosen algorithm, e.g., pIRS [8]. The Module may receive or not a genome reference entry. When no input is available, the reads are randomly generated, according to the algorithm specified in the instance of the Generator Module. Besides, it could also receive raw reads, to filter them in a quality trimming hot spot, or by modifying them for any purpose.

Genome Assembly: this is the module where the reads are assembled. It may output contigs, or scaffolds, which may be analyzed in the Assembly Evaluation Module, or may work as re-input to Genome Assembly, calling hot spots capable of scaffolding, gap-filling, etc. It works with distinct assemblers, file formats, and sequencing technologies.

Assembly Evaluation: we calculate the features qualifying the assemblies in the Assembly Evaluation Module, throughout the Feature Calculator instances. Not only the evaluation metrics but also some other post-analysis could be included here, such as genome comparisons. Those results, according to each application, may be still analyzed in the Statistics Module.

Statistics: the Statistics Module organizes/normalizes feature data, statistically testing, and/or visualizing data through the instances available in the Statistics Calculator Modules.

Generator: the Generator Module is an abstraction of service for the Reads Generation Module, where distinct instances may be present.

Assembler: the Assembler Module is the most important service we have; it abstracts assembler software. It varies in the way of assembling, how outputs, and how to receive input data.

Feature Calculator: is the abstraction for feature calculation and may be instantiated into classes capable of calculating single features. It may include more complex modules that calculate many features at once, such as QUAST.

Statistics Calculator: finally we have the module where all statistical tests and graphs are calculated and created.

2.3   Extending Abstract Classes

The greatest advantage of using a framework is the easy way in how new items are added. For example, consider in Fig. 2, the gray boxes as extensions of the abstract classes. The Framework Manual, provided on GitHub https://github.com/neumannguib/GAAF-Framework, explains that procedure to each abstract class. It is always the same: receive the abstract class as a parameter and define the abstract method and other methods required by specific applications.

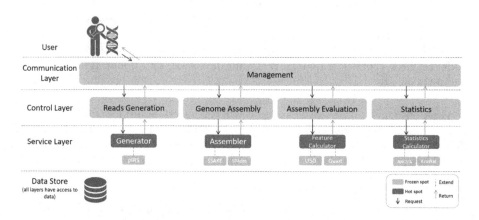

Fig. 2. Framework architecture extended

3   Main Contributions

The Framework contributes to the field in the way it is used and instantiated. In comparison to a simple script which calls tools, many options arise when framework approaches are considered. Not all modules need to be admitted into applications, and the abstraction available support easy adaptation to distinct workflows. Users' contributions can even add more possibilities to other applications, creating an excellent platform for Genome Assembly.

GAAF creates the possibility to reanalyze entire experiments, without running everything from scratch. Thus, when an error occurs or when the experiment execution is interrupted, the scientist may continue running the test from where it was in the process. Moreover, one may add a new assembler and reanalyze the experiment considering the assemblies from that assembler, in conjunction with the assemblies and features previously generated.

An additional benefit is a more effortless practice of adding new tools without previous knowledge about other GAAF components. It is just necessary to extend an abstract class and to create tool-specific functions. Besides, logging support, configuration files, and complete report functionality give users enough experiment details to be easily reproduced by others.

Future works include the development of a graphic interface, for a more straightforward Experiment Design, visualization of the process status, process manipulation, and data access. It is also vital to create a knowledge base, persisting experiments data, through a DBMS (Database Management System). It will reinforce the possibility to reanalyze and upgrade previous experiments.

References

1. Bradnam, K.R., et al.: Assemblathon 2: evaluating de novo methods of genome assembly in three vertebrate species. GigaScience **2**(1), 2047–217X (2013). https://doi.org/10.1186/2047-217X-2-10
2. Castro, C.J., Ng, T.F.F.: U50: a new metric for measuring assembly output based on non-overlapping target-specific contigs. J. Comput. Biol. **24**(11), 1071–1080 (2017). https://doi.org/10.1089/cmb.2017.0013
3. Bishop, C.M.: Pattern Recognition and Machine Learning. Springer, New York (2006)
4. Darling, A.E., Tritt, A., Eisen, J.A., Facciotti, M.T.: Mauve assembly metrics. Bioinformatics **27**(19), 2756–2757 (2011). https://doi.org/10.1093/bioinformatics/btr451
5. Denton, J.F., Lugo-Martinez, J., Tucker, A.E., Schrider, D.R., Warren, W.C., Hahn, M.W.: Extensive error in the number of genes inferred from draft genome assemblies. PLoS Comput. Biol. **10**(12), e1003998 (2014). https://doi.org/10.1371/journal.pcbi.1003998
6. Fayad, M., Schmidt, D.: Object-oriented application frameworks naval open systems architecture strategy view project unified software engines (USEs): a unified approach to building software systems View project. Commun. ACM (1997). https://doi.org/10.1145/262793.262798
7. Gurevich, A., Saveliev, V., Vyahhi, N., Tesler, G.: QUAST : quality assessment tool for genome assemblies. Bioinformatics **29**(8), 1072–1075 (2013). https://doi.org/10.1093/bioinformatics/btt086
8. Hu, X., et al.: pIRS: Profile-based Illumina pair-end reads simulator. Bioinformatics **28**(11), 1533–1535 (2012). https://doi.org/10.1093/bioinformatics/bts187
9. Hunt, M., Kikuchi, T., Sanders, M., Newbold, C., Berriman, M., Otto, T.D.: REAPR: a universal tool for genome assembly evaluation. Genome Biol. **14**(5), R47 (2013). https://doi.org/10.1186/gb-2013-14-5-r47
10. Land, M.L., et al.: Quality scores for 32,000 genomes. Stan. Genomic Sci. **9**(1), 20 (2014). https://doi.org/10.1186/1944-3277-9-20
11. Landin, N., Niklasson, A., Regnell, B.: Development of Object-Oriented Frameworks (1995)
12. Liu, D., Hunt, M., Tsai, I.J.: Inferring synteny between genome assemblies: a systematic evaluation. BMC Bioinform. **19**(1), 1–6 (2018). https://doi.org/10.1186/s12859-018-2026-4
13. Magoc, T., et al.: GAGE-B: an evaluation of genome assemblers for bacterial organisms. Bioinformatics (Oxford, England) **29**(14), 1718–1725 (2013). https://doi.org/10.1093/bioinformatics/btt273
14. Vezzi, F., Narzisi, G., Mishra, B.: Feature-by-feature - evaluating de novo sequence assembly. PLoS ONE **7**(2), e31002 (2012). https://doi.org/10.1371/journal.pone.0031002

15. Vezzi, F., Narzisi, G., Mishra, B.: Reevaluating assembly evaluations with feature response curves: GAGE and assemblathons. PLoS ONE **7**(12), e52210 (2012). https://doi.org/10.1371/journal.pone.0052210
16. Waterhouse, R.M., et al.: BUSCO applications from quality assessments to gene prediction and phylogenomics. Mol. Biol. Evol. **35**(3), 543–548 (2017). https://doi.org/10.1093/molbev/msx319

Identifying *Schistosoma mansoni* Essential Protein Candidates Based on Machine Learning

Francimary P. Garcia⬛, Gustavo Paiva Guedes⬛,
and Kele Teixeira Belloze$^{(\boxtimes)}$⬛

Federal Center for Technological Education of Rio de Janeiro (CEFET/RJ),
Rio de Janeiro, Brazil
francigarciaoliveira@gmail.com,
{gustavo.guedes,kele.belloze}@cefet-rj.br

Abstract. The essentiality of proteins is a valuable characteristic in research related to the development of new drugs. Neglected diseases profit from this characteristic in their research due to the lack of investments in the search for new drugs. Among the neglected diseases, we can highlight the schistosomiasis caused by the *Schistosoma mansoni* organism. This organism is a major cause of infections in humans and only one drug for its treatment is recommended by the World Health Organization. This fact raises a concern about the development of drug resistance by this organism. In this context, the present work aims to identify *S. mansoni* essential protein candidates. The methodology uses a machine learning approach and makes use of the knowledge of protein essentiality characteristics of model organisms. Experimental results show the Random Forest algorithm achieved the best performance in predicting the protein essentiality characteristic of *S. mansoni* compared to the other evaluated algorithms.

Keywords: Machine learning · Essentiality of protein · *S. mansoni*

1 Introduction

Schistosomiasis is a neglected disease caused by helminth organisms (parasitic worms) of the genus Schistosoma. Among these organisms, *Schistosoma mansoni* is a major cause of infections in humans [12]. Because it is an endemic disease in underdeveloped countries such as Brazil, there is no investment by the pharmaceutical industry in new drugs. For its treatment, the World Health Organization (WHO) recommended only one drug (praziquantel) which has been used in clinical practice for almost four decades [9]. Due to the high incidence of reinfection in endemic areas, there is concern about the development of drug resistance by this helminth. This fact justifies the occurrence of researches that seeks the identification of characteristics in the *S. mansoni* genes and proteins to enable the search for new drugs. Essentiality is a relevant characteristic because it refers

© Springer Nature Switzerland AG 2020
L. Kowada and D. de Oliveira (Eds.): BSB 2019, LNBI 11347, pp. 123–128, 2020.
https://doi.org/10.1007/978-3-030-46417-2_12

to the essential genes and proteins of an organism, which are indispensable for sustaining cell life [13].

Essential genes or proteins identification has been the subject of several studies in the computing field. Methods based on homology, evolution analysis, and machine learning stand out [10]. Machine learning algorithms use trained data to discover underlying patterns, build models, and make predictions based on the best-fit model [7]. In the literature, several studies use this approach to make predictions about the essentiality of genes and proteins for either eukaryotic or prokaryotic organisms [14]. Among these, some studies show that characteristics based on the primary protein sequence, such as the frequency of amino acid occurrence, are suitable for machine learning tasks [6].

Given the scenario presented, this paper aims to use a machine learning approach to identify *S. mansoni* proteins with essentiality characteristic. To this end, this work draws on knowledge of the essential and non-essential proteins of model organisms. The methodology for this work includes the integration of the primary sequences of the proteins of the model organisms labeled with essentiality or non-essentiality information. After that, the machine learning algorithms are applied to the training task and subsequent prediction of the essentiality characteristic of the organism proteins. The results present the evaluation measures of the algorithms, besides the number of essential and non-essential proteins identified in the study organism. Moreover, the paper presents a comparative analysis of the prediction results and the results obtained in [4]. The objective of the referred paper is to search for *S. mansoni* drug targets, focusing on the identification of protein essentiality characteristics using a homology-based approach.

Besides this introduction, the remainder of this paper is organized as follows. Section 2 presents the methodology used to accomplish this research. Section 3 presents the results. Section 4 describes the final remarks.

2   Methodology

This work proposes the identification of *S. mansoni* proteins with essentiality characteristics. The following macro activities are intended to conduct this work: (i) data collection and integration, (ii) data transformation and (iii) classification and prediction of the essentiality characteristic.

2.1   Data Collection and Integration

The development of this work is based on the knowledge of three model organisms, namely: *Mus musculus*, *Caenorhabditis elegans*, and *Saccharomyces cerevisiae*. The proteome of each organism in fasta format was downloaded from the Ensembl database [5]. Essentiality and non-essentiality information (on experimentally tested proteins) from these organisms were downloaded from the Online Gene Essentiality (OGEE) database [2]. The proteome data is cleaned and integrated through a script developed for this purpose, and then the essentiality or non-essentiality information is added to each protein.

2.2 Data Transformation

Since the integrated data set refers to primary protein sequences, data transformation is required to enable the use of machine learning algorithms. Thus, the data are transformed into a discrete protein model, in which normalized occurrence frequencies of the 20 native amino acids in each protein are calculated, called Amino Acid Composition (AAC) [8]. The iLearn[1] platform is used to obtain the AAC values. As a result, a consolidated data set is created with the labeled AAC information of the proteins of the model organisms. This data set, generated in the ARFF (Attribute-Relation File Format) format, is organized into 21 attributes, where 20 of them refer to each amino acid, and one refers to the essential or non-essential label of each protein.

2.3 Classification and Prediction

Random Forest (RF), J48, SVO, and Logistic algorithms are employed to create the classification model for the transformed data set according to the AAC characteristic. All the experiments use the 10-fold cross-validation technique. Then, each classification model is used to predict the essentiality characteristic of the proteins of the study organism. At the end of the process, a list containing all study organism proteins is generated, where each protein is labeled as essential or non-essential.

The Weka tool [3] performs these tasks. The tool Experimenter functionality was used for choosing classification algorithms. To this end, the training data set is loaded and tested according to a list of different algorithms. Algorithms that presented the best results for the precision measurements were selected.

3 Results

This section presents the results obtained according to the methodology showed. Initially, an experiment was conducted to train the four classification algorithms mentioned (RF, J48, SVO, and Logistic) and compare their results. Subsequently, based on the algorithm that presented the best performance, the essentiality characteristic was predicted on the unclassified *S. mansoni* proteins. The *S. mansoni* proteome in fasta format was downloaded from the Ensembl database.

3.1 Training of Classification Algorithms

In this experiment, the transformed training data set consisted of the essential and non-essential proteins of the three model organisms. Thus, the data set consisted of 25,599 proteins, of which 6,149 were essential, and 19,450 were non-essential proteins. This data set was transformed according to the AAC characteristic and submitted to execution in the four classification algorithms.

---

[1] http://ilearn.erc.monash.edu/.

Due to the imbalance between the number of essential and non-essential proteins of the training data set, the results presented a biased behavior in the initial execution. Thus, it was necessary to apply techniques to perform data balancing. In this study, we used the oversampling and undersampling techniques, and then a new execution was conducted with each one. For the oversampling balancing scenario, the Synthetic Minority Oversampling TEchnique (SMOTE) filter with a rate of 215% was used. For the undersampling scenario, the SpreadSubSample filter with distributionSpread = 1 was defined. Both filters are present in the Weka tool. Table 1 presents the results of this task for each algorithm.

Table 1. Performance comparison of different classification algorithms using the oversampling and undersampling techniques.

		RF	J48	Logistic	SMO
Oversampling	Accuracy	0.791	0.708	0.669	0.676
	AUC-ROC	0.873	0.722	0.718	0.676
Undersampling	Accuracy	0.681	0.651	0.664	0.666
	AUC-ROC	0.751	0.672	0.709	0.666

The performance of the classification algorithms was compared using the accuracy and AUC-ROC curve measurements. The accuracy corresponds to the ratio of instances correctly classified by the algorithms. This value is the ratio of the sum of the confusion matrix occurrence diagonals (True Positive of Class A + True Positive of Class B) related to the number of instances [1]. The AUC (Area Under The Curve) ROC (Receiver Operating Characteristics) curve provides a comparison of predicted and expected target values in a classification [11]. In this sense, the Random Forest algorithm presented the best results, considering each data balancing technique. However, the accuracy and ROC curve percentages are higher for oversampling.

3.2   Prediction of *S. mansoni* Proteins

To perform the prediction task, we used the unclassified *S. mansoni* protein data set, consisting of 11,774 proteins. Initially, the data set was transformed according to the AAC characteristic. Thus, we used the classification model generated by the Random Forest algorithm with oversampling balancing (described in Sect. 3.1) to predict the essentiality characteristic of the data set proteins. The prediction classified 1,412 proteins as essential and 10,362 as non-essential.

The result of this prediction was compared to the result found in the work of Garcia and Belloze [4]. This work uses a homology-based approach and identifies *S. mansoni* proteins that are orthologous to the essential proteins of the three model organisms: *Mus musculus*, *Caenorhabditis elegans* and *Saccharomyces cerevisiae*. The work indicates 138 essential protein candidates of *S. mansoni*.

Finally, it presents a list of the ten best-ranked proteins of *S. mansoni*, considering not only the essentiality characteristic but also the protein druggability.

Compared to the approach proposed in this paper (i.e., machine learning-based), of the 1,412 proteins classified as essential, 52 were also indicated as essential in the work proposed by Garcia and Belloze [4]. When compared with the ten best-ranked proteins, six were reported as essential (by both approaches) and can be seen in Table 2. The table shows the identifier and description of the *S. mansoni* protein, the protein prediction probability obtained in the machine learning-based approach, and the ranking among the ten indicated in the homology-based approach.

Table 2. *S. mansoni* essential protein candidates according to homology-based and machine learning approaches.

Protein ID	Description	Prediction	Rank
Smp0466001	Actin-1	0.93	5
Smp1837101	Putative actin	0.91	4
Smp1619201	Putative actin	0.82	6
Smp0182401	Cell division control protein 48 aaa family protein	0.63	9
Smp0182402	Cell division control protein 48 aaa family protein	0.66	8
Smp2031301	Putative uncharacterized protein	0.54	3

4   Final Remarks

The importance of identifying the genes and proteins essentiality characteristic, associated with the vast availability of biological data, has stimulated research using computational approaches. Such research is important to support the process of developing new drugs and particularly for neglected diseases, due to the low investment of the pharmaceutical industry in this class of diseases.

The methodology proposed in this paper uses a machine learning-based approach to predict essential protein candidates of *S. mansoni*, which causes schistosomiasis, a significant neglected disease. In this context, a characteristic based on the proteins primary sequence of model organisms, called Amino Acid Composition, is used. The experimental results point to the Random Forest algorithm, considering the balance of data set by oversampling, with better performance in predicting the essentiality characteristic of *S. mansoni* proteins. Proteins classified as essential candidates for *S. mansoni* may support more specific experimental research into the discovery of new drug targets in the fight against schistosomiasis.

As an evolution of this research, it is suggested the inclusion of other features related to the proteins primary sequence in order to enrich the classification model generated by the machine learning approach. Examples of features include the frequency of amino acid pairs of a specific size and descriptors that characterize amino acids according to their physicochemical properties. In other future work, it is intended to use data from other model organisms for the training task and to compare with the results of the present work. Also, for validation, it is meant to use other organisms with labeled essentiality characteristics.

References

1. Arora, R.: Comparative analysis of classification algorithms on different datasets using weka. Int. J. Comput. Appl. **54**(13), 0975–8887 (2012)
2. Chen, W.H., Minguez, P., Lercher, M.J., Bork, P.: OGEE: an online gene essentiality database. Nucleic Acids Res. **40**(D1), D901–D906 (2011)
3. Frank, E., Hall, M.A., Witten, I.H.: The Weka Workbench. online appendix. In: Frank, E., Hall, M.A., Witten, I.H. (eds.) Data Mining Practical Machine Learning Tools and Techniques. Morgan Kaufmann, Cambridge (2016)
4. Garcia, F.P., Belloze, K.T.: Integração de dados na detecção de alvos para fármacos de schistosoma mansoni. In: Anais do XII Brazilian e-Science Workshop. SBC (2018)
5. Kersey, P.J., et al.: Ensembl genomes 2018: an integrated omics infrastructure for non-vertebrate species. Nucleic Acids Res. **46**(D1), D802–D808 (2017)
6. Liu, X., Wang, B.J., Xu, L., Tang, H.L., Xu, G.Q.: Selection of key sequence-based features for prediction of essential genes in 31 diverse bacterial species. PLoS ONE **12**(3), e0174638 (2017)
7. Min, S., Lee, B., Yoon, S.: Deep learning in bioinformatics. Brief. Bioinf. **18**(5), 851–869 (2017)
8. Nakashima, H., Nishikawa, K., Ooi, T.: The folding type of a protein is relevant to the amino acid composition. J. Biochem. **99**(1), 153–162 (1986)
9. Neves, B.J., et al.: Discovery of new anti-schistosomal hits by integration of QSAR-based virtual screening and high content screening. J. Med. Chem. **59**(15), 7075–7088 (2016)
10. Peng, C., Lin, Y., Luo, H., Gao, F.: A comprehensive overview of online resources to identify and predict bacterial essential genes. Front. Microbiol. **8**, 2331 (2017)
11. Sahoo, G., Kumar, Y.: Analysis of parametric & non parametric classifiers for classification technique using weka. Int. J. Inf. Technol. Comput. Sci. (IJITCS) **4**(7), 43 (2012)
12. WHO: Shistosomiasis (2019). http://www.who.int/schistosomiasis/en/
13. Zhang, R., Ou, H.Y., Zhang, C.T.: Deg: a database of essential genes. Nucleic Acids Res. **32**(suppl-1), D271–D272 (2004)
14. Zhang, X., Acencio, M.L., Lemke, N.: Predicting essential genes and proteins based on machine learning and network topological features: a comprehensive review. Front. Physiol. **7**, 75 (2016)

Author Index